Scientific & Technological Literacy

TESTING FOR EFFECTS OF CHEMICALS ON ECOSYSTEMS

A Report by the Committee to Review
Methods for Ecotoxicology

Commission on Natural Resources

National Research Council

NATIONAL ACADEMY PRESS
Washington, D.C.   1981

NOTICE: The project that is the subject of this report was approved by the Governing Board of the National Research Council, whose members are drawn from the Councils of the National Academy of Sciences, the National Academy of Engineering, and the Institute of Medicine. The members of the Committee responsible for the report were chosen for their special competences and with regard for appropriate balance.

This report has been reviewed by a group other than the authors according to procedures approved by a Report Review Committee consisting of members of the National Academy of Sciences, the National Academy of Engineering, and the Institute of Medicine.

This study was supported by the Council on Environmental Quality, Contract No. EQ-9AC-008.

The National Research Council was established by the National Academy of Sciences in 1916 to associate the broad community of science and technology with the Academy's purposes of furthering knowledge and of advising the federal government. The Council operates in accordance with general policies determined by the Academy under the authority of its congressional charter of 1863, which establishes the Academy as a private, nonprofit, self-governing membership corporation. The Council has become the principal operating agency of both the National Academy of Sciences and the National Academy of Engineering in the conduct of their services to the government, the public, and the scientific and engineering communities. It is administered jointly by both Academies and the Institute of Medicine. The National Academy of Engineering and the Institute of Medicine were established in 1964 and 1970, respectively, under the charter of the National Academy of Sciences.

Library of Congress Cataloging in Publication Data

Environmental Studies Board. Committee to Review
  Methods for Ecotoxicology.
  Testing for effects of chemicals on ecosystems.

  Includes bibliographical references.
  1. Chemicals—Environmental aspects. 2. Toxicity testing. 3. Environmental monitoring. I. Title.
QH545.A1E59   1981   574.5'222   81-38392
ISBN 0-309-03142-7   AACR2

Available from

NATIONAL ACADEMY PRESS
National Academy of Sciences
2101 Constitution Avenue, N.W.
Washington, D.C. 20418

Printed in the United States of America

Contents

Preface

An assessment of the potential extent of environmental hazards posed by intentional or accidental release of chemicals requires the collection of a complex set of data. These data must describe a variety of characteristics indicative of multi-species interactions within ecosystems. Recognizing that most current assessments rely primarily on data generated by single-species tests, the Council on Environmental Quality (CEQ) and the U.S. Environmental Protection Agency (EPA) initiated a review of available laboratory test systems and data evaluation schemes for predicting effects of chemicals on ecosystems. The Environmental Sciences Division of Oak Ridge National Laboratories was asked to conduct this review. In addition, the National Research Council (NRC) was asked to assist in structuring the review and to address the problem of advancing the role of applied ecology in toxicological assessments. The purpose of these activities was to assist EPA in improving the quality of scientific information for making decisions and in developing environmental risk assessments that are needed to implement the Toxic Substances Control Act.

In the fall of 1979, the Environmental Studies Board was given responsibility for the NRC study and appointed the Committee to Review Methods for Ecotoxicology. The charge to the Committee was to identify characteristics of ecological systems that would indicate hazardous effects of chemicals beyond the level of single species (i.e., effects on interactions among populations as well as on the structure and functional processes of ecosystems), to establish criteria for suitable testing schemes, and to evaluate the effectiveness of available test systems in assessing effects of chemicals within ecosystems.

During the course of this study, the Committee benefited greatly from discussions with and contributions by individuals from universities, industries, and governmental agencies. We are particularly grateful to J. Sprague, University of Guelph; H.L. Ragsdale, Emory University; G.W. Salt, University of California, Davis; R.H. Kadlec, University of Michigan; L.A. Norris, U.S. Forest Service; P.J. McCall, D.A. Laskowski, R.L. Swann, and H.J. Dishburger, Dow Chemical, USA; J. Berg, University of Tennessee; H.T. Band, Michigan State University; R.L. Lowe, Bowling Green State University; R.R. Lassiter, Environmental Research Laboratory; W.R. Swain, Large Lakes Research Laboratory; P.H. Gleick, University of California, Berkeley; and R.A. Schneider, University of California, Berkeley. Many of the materials provided by these individuals are published in a supplemental volume to this report. The Committee also appreciates the assistance given by J.D. Buffington, CEQ, and by J.V. Nabholtz and J.J. Reisa, both of EPA.

On behalf of the Committee, I would like to thank Suellen Pirages, Elizabeth Panos, and Lawrence Wallace of the National Research Council for their contributions in preparing this report. Other staff members providing assistance include Janis Horwitz, Catherine Iino, Raphael Kasper, Estelle Miller, Christina Shipman, and Robert Rooney.

Finally, I wish to express my gratitude to each member of the Committee for his efforts toward the successful completion of the study.

JOHN CAIRNS, JR., *Chairman*
Committee to Review Methods for Ecotoxicology

Executive Summary

Examination of available test systems and evaluation strategies makes it evident that the current basis for developing such strategies is inappropriate. The underlying question should concern the types of information that must be acquired to determine potential environmental hazard of new chemicals, not the tests that are available to provide data rapidly and conveniently. Once the needed information is identified, attention should be given to developing test systems that will provide it in the most effective and economical manner. This report stresses that current modes of testing (i.e., the predominant reliance on single-species tests and progression from simple acute toxicity tests to complex chronic toxicity tests) must be redesigned to provide a variety of data about (a) responses of single species, (b) impacts on interactions among different species, and (c) enhancement or impairment of functional processes within different types of ecosystems.

The vulnerability of a system to the presence of a chemical will depend on many factors, including the chemical, physical, and biological properties of the ecosystem, as well as the characteristics and modes of entry of the chemical. Because of these factors, evaluations of impact cannot be made solely on the basis of data generated by single-species tests. As discussed in Chapter 1, limitations of single-species tests include the following.

(1) Current laboratory tests examine only the responses of individuals, which are then averaged to give a mean response for the test species.

(2)   With given constraints of limited finances and number of personnel, it is not possible to identify the most sensitive species or group of species.

(3)   The data are too limited in scope for extrapolations to be made from them of responses of other (even closely related) species.

(4)   Indirect effects resulting from population or species interactions cannot be observed.

(5)   Conditions within which single-species tests are performed lack the realism of natural habitats.

This report should not be interpreted as a criticism of single-species testing. In several sections, the report notes that single-species tests are essential to evaluations of hazard caused by chemicals introduced into the environment. Single-species toxicity testing is an appropriate way to determine toxicological effects on lethality, growth, reproductive success, behavior, and a variety of other individual characteristics. The report is critical, however, of using single-species data to predict effects of chemicals upon interactions within and among species (e.g., competition, predation, and relationships between host and parasite) and upon effects at the system level (e.g., alterations in flow of energy, nutrient spiraling, and diversity).

Single-species tests, if appropriately conducted, have a place in evaluating a number of phenomena affecting an ecosystem. However, they would be of greatest value if used in combination with tests that can provide data on population interactions and ecosystem processes.

Chapter 2 discusses factors that determine the fate of a chemical within an ecosystem. Chemical and physical properties of a compound influence its movement through environmental media, as do properties of biotic and abiotic components of an ecosystem.

Partition coefficients should be determined for the movement of a chemical across the interface between environmental media and for movement within the structural network of a particular ecosystem.

Many processes contribute to the reduction or increase of potential toxicity of a chemical. These can be abiotic processes (e.g., photolysis) or biotic transformation (e.g., microbial degradation). Numerous transformation products can be generated within the ecosystem; thus, particular substrate characteristics and the identity of transformation products should be considered.

The toxic potential of all major transformation or degradation products of the parent compound should be identified.

Variability occurs at many points beginning at the source of discharge of a substance and ending with variances in individual responses to its presence. Because of this variability the test conditions should be carefully designed and the data interpreted with caution.

The distribution and subsequent fate of chemicals in the environment determine the dose delivered to various individual biotic components. Because the dose, in combination with the duration of exposure, is of prime importance for determining whether an adverse effect will occur, concentration and exposure time must be examined carefully. The most promising avenues for obtaining this information early in an assessment process will be found in the application of fugacity or partitioning equations combined with rates of degradation, transformation, and transfer.

Test conditions should be carefully designed to account for variations in natural systems that will affect dose delivered to the biota as well as exposure time in any particular compartment.

Certain characteristics of populations and ecosystems should be considered in designing test systems to evaluate potential impacts of chemicals. These characteristics are discussed in Chapter 3. Pertinent characteristics of populations include changes in age distribution, mortality, fecundity, growth rates, migratory behavior, phenotypic variation, and mutation rates. Interactions of stress factors, and changes in behavior and spatial relationships are attributes of individual organisms, but at some future time may be described for ecosystems as well. Methods should be developed for detecting change in such properties related to systems as diversity, productivity and biomass, connectivity, resistance and resilience, and genotypic or taxonomic variability.

Certain factors of an ecosystem can influence the impact of a chemical, and these factors must be considered in evaluating the potential hazards of substances. Such factors include the capacity of the system to store or detoxify the chemical, adaptive potential of species, species distribution and density within the ecosystem, and climatic changes over time and space.

Research and development should be directed towards designing and validating test systems and procedures that will detect changes in ecosystem and population attributes. Because natural fluctuations can mask changes resulting from the presence of a chemical, methods should be developed to distinguish natural variations from changes made by chemicals.

Any valid assessment strategy must include a set of well-defined criteria for evaluating test procedures, as discussed in Chapter 4. These include the ability to predict effects over a broad range of test conditions, to produce verifiable data, to be easily replicated in several laboratories, and to analyze the data using accepted statistical techniques. The test systems should be as environmentally realistic as possible by duplicating the natural habitat of test species as well as the form and potential fate of the test chemical. The most critical of these criteria are standardization, realism, and the capability of verifying test procedures. These three also require the greatest effort in test development.

No one type of test can provide sufficient information to make accurate predictions of chemical impacts on the environment. An integrated strategy should require tests of effects on individuals of a single species, interactions within populations, interactions among species, and effects on structure and function of ecosystems for terrestrial and aquatic systems. Several test systems are currently available for use in evaluation schemes as discussed in Chapter 4. Single-species tests are of considerable value in establishing suitable dose ranges for use in multi-species tests. Most test systems currently used test not population responses to a chemical but responses of a few individuals of a particular species. Measurements of changes in population dynamics exist and can be developed into suitable test procedures. Short-term effects of chemicals on functional processes of ecosystems (e.g., mineralization or nutrient cycles) can be tested using multi-species microcosms.

Further research should be directed toward verifying the realism of microcosm systems. Continued development is needed to enable the use of these systems for long-term studies and for assessment of hazards to larger organisms than are being used currently.

The final chapter of this report discusses in detail an appropriate assessment strategy for an integrated multi-level approach to the collection of data. Before a decision can be made about the potential hazard of any new chemical, however, four classes of information should be gathered.

1. *Characterization of test substance*: information on chemical and physical properties, estimates of fate within and among ecosystems, and estimates of dose and exposure time for biotic components of an ecosystem.

2. *Physiological responses of species*: data on individuals of representative species indicating morphological, biochemical, genetic, and pathological changes related to the presence of the chemical.

3. *Multi-species responses*: information on changes in interactions among organisms, including changes in population or system structure and changes in patterns of interaction among species (e.g., predation, competition, and migratory behavior patterns).

4. *Ecosystem responses*: data on changes in functional processes that affect the resistance and resilience of the ecosystem.

Research should be conducted to develop test procedures that can provide multiple sets of data (e.g., data on physiological responses as well as on the interactions among species). Tests should be designed to provide short-term results about long-term effects.

Impacts of chemicals above the level of individuals can be detected only if natural conditions of ecosystems are well documented. Characterizations of the structure and functional processes of general ecosystem types as well as of specific ecosystems must be developed independently of concerns about the impact of any particular chemical. This report recommends establishing baseline ecosystem studies to make possible a complete understanding of natural systems.

The report also recommends a design for an assessment strategy along with the action needed to implement it. The design includes baseline ecosystem studies, integrated laboratory testing schemes and mathematical models of ecosystem dynamics. The proposed strategy should allow detection of even low probability impacts by combining results of many different tests and models. It should be emphasized that as more knowledge and experience accumulates, it may be feasible to reduce the number of laboratory tests. It is hoped that, as the process evolves, it will become possible to obtain more information from fewer experiments.

Because of the lack of individuals trained in ecotoxicology, implementation of many of the recommendations, particularly those calling for more research, will be difficult.

To remedy this situation, more funds should be made available to support the training of students in ecotoxicology.

Introduction

Under the terms of the Toxic Substances Control Act (TSCA), the Administrator of the U.S. Environmental Protection Agency (EPA) may regulate the manufacture, distribution in commerce, use, and disposal of chemical substances and mixtures proven to present an unreasonable risk to human health or the environment (U.S. Congress 1976). The Act authorizes the Administrator to require an assessment of the magnitude of expected exposures and the potential effects of new chemicals or of old chemicals used in new ways. In response to this legislation, EPA has discussed proposing guidelines that describe the type of data needed for the assessment of new chemicals (U.S. EPA 1979).

Current test methods used to generate toxicological data, however, are based on the use of individual organisms of a single species. The choice of laboratory species to use in tests has depended on either the similarity of selected organisms to human biological systems or the ease of rearing and testing these organisms in the laboratory. Minimal attention has been given to their relevance for use in an assessment of chemical effects on ecosystems (Norris 1980). Thus, few evaluation schemes exist that provide an adequate data base with which to estimate potential hazards to ecosystems.

An ecosystem is a complex of both biotic and abiotic components; it is at least partially self-sustaining and self-regulating. The biotic components consist of plants, animals, and microorganisms that form a network at specific geographic locations. The abiotic components include the physical environment (air, water, soil, and sediment) and other nonbiological material within the geographic boundaries of the system.

1

An ecosystem is not merely a grouping of species at some specific site; it is a product of complex interactions between and among living and nonliving components. The interactions have identifiable structures and functions. For example, ecosystems have well-defined internal spatial patterns that reflect the response of component species to differences in local environments and the responses resulting from competition between and within species. Biotic and abiotic relationships also exist; for example, the transfer of energy through a distinct pathway that is determined by the energy requirements and feeding relationships of species within a particular ecosystem. Levels of productivity in an ecosystem are the result of many processes that depend on interactions among such system components as nutrient cycles, predation, and reproduction.

If a prediction about the behavior of an ecosystem is to be made with some accuracy, the parameters describing the interrelationships among component parts of a system must be well defined. Because the behavior of an ecosystem is more frequently determined by the way in which certain processes change in response to a disturbance than by alterations in species composition, a complete understanding of the structure of a system may not be as important as an adequate understanding of functional processes. The combination of species present in an ecosystem can and often does change along gradients of space and time, but as long as the functional processes continue, the system remains viable. In addition, although many structural and functional changes can occur in response to a disturbance, the magnitude of the change within the ecosystem often is not directly proportional to the extent of the disturbance; that is, the responses of ecosystems to stress are generally nonlinear.

Because ecosystems are complex, simple cause-and-effect relationships between the introduction of a chemical and the response of the system are difficult to detect in most instances. This difficulty complicates the achievement of the type of assessment required by TSCA. Furthermore, a chemical introduced into similar ecosystems may not produce the same type of change in each. Variations in the physical environment (e.g., temperature or moisture) and small differences in species composition can change the fate of the chemical (e.g., storage or degradation), leading to different impacts within the system. The magnitude of the impact also is determined by inherent system qualities of resistance or resilience and by natural fluctuations in system processes.

The task of the Administrator of EPA in carrying out the provisions of TSCA is made more difficult by the overwhelming reliance on data generated from single-species tests. It is a general practice of ecotoxicologists to perform multi-species and ecosystem tests only after single-species tests have been conducted on a standard set of organisms (NRC

1975, Conservation Foundation 1978, Cairns et al. 1978, Dickson et al. 1979). This practice may have arisen as a result of the historical development of the field of toxicology and the lack of emphasis on development of population and ecosystem level tests. There is, however, no scientific justification for continuing with sequential testing strategies; *single-species, population, and system-level tests furnish different types of information, all of which are needed before sound judgments can be made about the potential environmental hazard of any chemical.*

In the course of this study several questions began to emerge regarding the broader issue of an appropriate strategy for collection of data. These questions provided the foci for the chapters that follow.

Chapter 1: Are single-species tests sufficient for assessing the entire spectrum of change produced in an ecosystem by the introduction of chemical substances?

Chapter 2: What chemical, physical, and biological factors influence the fate of a chemical within an ecosystem?

Chapter 3: Which properties of populations and ecosystems are most likely to be affected by the presence of chemicals, and which of these are likely to influence the impact of a chemical?

Chapter 4: What are the adequacies and deficiencies of current test systems in providing data with which chemical impacts on populations and ecosystems can be assessed?

Chapter 5: What assessment strategy would best assure accurate evaluation of chemical effects on ecosystems?

Within this framework the report reviews the type of information that is needed for predicting the effects of chemicals on ecosystems, and:

• identifies those characteristics of populations and ecosystems that might serve in assessing adverse effects of chemicals above the level of single species;

• recommends criteria for selecting test procedures;

• evaluates the effectiveness of available test systems, drawing, in part, upon a recent literature survey (Hammons 1980); and

• recommends a strategy for evaluating effects of chemicals on ecosystems.

REFERENCES

Cairns, J., Jr., K.L. Dickson, and A.W. Maki, Eds. (1978) Estimating the hazard of chemical substances to aquatic life. ASTM Special Technical Publication 657. Philadelphia, Pa.: American Society for Testing and Materials.

Conservation Foundation (1978) Approaches for developing testing guidelines under the Toxic Substances Control Act. Washington, D.C.: Conservation Foundation.

Dickson, K.L., A.W. Maki, and J. Cairns, Jr., Eds. (1979) Analyzing the hazard evaluation process. Proceedings of a Workshop held in Waterville Valley, New Hampshire, August 14-18, 1978. Washington, D.C.: American Fisheries Society, Water Quality Section.

Hammons, A.S., Ed. (1980) Methods for ecological toxicology: a critical review of laboratory multispecies tests. EPA-560/11-80-026. Oak Ridge, Tenn.: Oak Ridge National Laboratories.

National Research Council (1975) Principles for evaluating chemicals in the environment. Committee for the Working Conference on Principles of Protocols for Evaluating Chemicals in the Environment. Environmental Studies Board and the Committee on Toxicology. Washington, D.C.: National Academy of Sciences.

Norris, L.A. (1980) Ecotoxicology at the watershed level. *In* Working Papers for the Committee to Review Methods for Ecotoxicology. Available in limited supply from the Environmental Studies Board, Commission on Natural Resources. Washington, D.C.: National Academy Press.

U.S. Congress (1976) Toxic Substances Control Act. Public Law 94-469.

U.S. Environmental Protection Agency (1979) Toxic substance control: discussion of premanufacture testing policy and technical issues. Federal Register 44(53):16240-16292.

1 Assessment of Chemical Toxicity

Evaluating the potential environmental impact of a chemical is a difficult but not impossible task. The natural environment consists of categories of ecosystems (e.g., terrestrial, freshwater, or marine) within which many diverse combinations of biotic and abiotic components exist. Attempts to understand the response of a system to any given substance are complicated by the diversity of physical, chemical, and biological factors and their interrelationships. Evaluation is further complicated by the potential for adaptation inherent in the biotic components, by the extent of diversity within an ecosystem, and by the range of differences in the magnitude of the responses by component parts.

The vulnerability of an ecosystem to disturbances depends upon many factors. They include: (a) properties of the ecosystem that contribute to its ability to either resist substantial changes resulting from the presence of a chemical or to return to the original state of the system after the chemical has been removed; (b) properties of the chemical and any transformation products; (c) type of exposure, e.g., acute or chronic, intermittent or continuous; (d) geographic location of the ecosystem relative to the point of release of the chemical; and (e) concentration of the chemical at the point of impact and amounts of the chemical moving through the ecosystem.

If the chemical is persistent and hence not subject to physical or biological degradation, impacts can occur at several points in the ecosystem. The chemical may inhibit basic physical or chemical mechanisms within the system; for example, it may limit the action of reducing agents

5

within the soil. The initial impact may be felt at the lowest level of the food chain, limiting productivity of the primary producers within the system. The material then could be transported and accumulated to biologically hazardous concentrations in the upper levels of a food chain. Sensitive functions may be the point of impact for the substance; for example, a sensitive mechanism in the nutrient cycle or the reproductive process of a key species may be affected.

USE OF SINGLE-SPECIES TESTS

Data generated from single-species tests have often been relied upon to estimate the concentrations of chemicals that, if released into the environment, would be incompatible with the orderly functioning of ecosystems. Single-species tests can provide much information on the concentrations and durations of exposures to chemicals that result in changes in survival, reproduction, physiology, biochemistry, and behavior of individuals within particular species, but results from such tests cannot predict or be used to evaluate aspects of chemical impacts beyond this level of biological organization.

Current assessment strategies focus on discrete biological populations found within broad ecosystem categories (e.g., terrestrial or aquatic). When tests are conducted that use individual species representative of these broad classes (e.g., EPA's discussion of proposed guidelines suggest rainbow trout and invertebrates for freshwater systems; quail for terrestrial systems; dogs, rats, and mice for mammalian responses), the results provide a means to determine both general and detailed toxic effects of specific chemicals on particular organisms. A range of chemical concentrations can be presented to groups of different test species and individual responses to each concentration can be observed. After dose-response curves are constructed using these test results, a general estimate of the extent of likely physiological response can be made for given environmental concentrations and expected duration of exposure. The popularity of such tests results from the quick identification of a dose and a corresponding effect.

Single-species tests range from tests of acute effects, where the major concern is rapid mortality, to highly sophisticated tests of chronic effects. The types of observations possible in chronic toxicity tests include long-term survival rates; growth rates; changes in reproduction; pharmacokinetic responses; mechanism of toxicity; pathological, biochemical, and physiological changes; and mutagenic, teratogenic, and carcinogenic rates. Current practices for ecotoxicological assessments, however, do not usually include much beyond observing changes in rates of survival, growth, and reproduction. Although acute tests are relatively simple and inexpensive, chronic tests are exceedingly complex and demand large commitments of finances, personnel and other resources. The types of data generated by chronic tests, however, are important for assessing the effects of chemicals on ecosystems, and further development of appropriate procedures for terrestrial and aquatic systems are needed.

LIMITATIONS OF SINGLE-SPECIES TESTS

Blanck et al. (1978) introduced two concepts that are relevant to a discussion of the limitations of single-species tests: ecological and pollutant realism. Test conditions are considered to be ecologically realistic if they reflect important characteristics of the natural environment, either for individual species or for ecosystems. Because single-species tests cannot delineate effectively the complex nature of ecosystem structure and functional processes, they lack ecological realism. A recent study of long-term effects of toxic substances on aquatic plant communities stressed that acute toxicity results are not adequate for making realistic predictions about effects of pollutants on natural systems (Hunding and Lange 1978). Sophisticated chronic, single-species tests can provide useful information for establishing effects for a given species under given conditions. But an uncritical application of these data to more general ecosystem conditions often leads to incorrect conclusions regarding the potential impact of a chemical.

One aspect of current testing practices is to attempt to identify the most sensitive species; this is difficult to determine with any type of test system but is particularly difficult by analysis of data from single-species tests alone. If detailed chronic toxicity tests were done for *all* species within a particular system, then the sensitive species probably could be identified, although this approach is highly impractical. A multi-species model of an ecosystem might permit identification of the more sensitive species, depending on inclusion of several factors in the model: a significant number of species representing the degree of diversity found in the ecosystem, detailed observations on physiological and behavioral responses for individual species, and a time period similar to the duration of expected chemical exposure in the ecosystem. As the number of species incorporated increases, the physical size and complexity of the test system also increases, resulting in another clearly impractical approach. However, comparative toxicity analyses using several multi-species systems (e.g., representative of competitive interactions, predator-prey relationships, or functional groups such as primary producers), could provide information on the more sensitive species within a particular ecosystem.

If a laboratory were provided with all of the needed resources (equipment, skilled technicians, and necessary funds) to develop and conduct the best-designed and most-detailed single-species test, the data generated would indicate only how a particular species might respond under a variety of conditions and in isolation from other species. Because organisms in natural settings are constrained in a number of ways that cannot be duplicated in single-species tests, results from such tests might lead to inaccurate predictions of effects. For example, constraints are imposed by inter- and intraspecies competition for nutrients and for suitable habitat or light; by the presence of predators and parasites; and by a host of other factors that operate simultaneously to prevent or expedite the process by which individuals of a species can react to changes in environmental conditions. The susceptibility of blue-green algae to DDT illustrates this point (Batterton et al. 1972). At high concentrations of NaCl the algae become extremely sensitive to low levels of DDT, but this sensitivity is not detected under normal environmental levels of NaCl.

Because certain natural stresses, such as predation and competition, are absent in laboratory single-species systems, an organism tested in these systems cannot be expected to respond to the chemical in the same way that it would in the natural habitat. Concentrations of 1 μg/kg of PCBs or 10 μg/kg of DDT do not produce any effects on pure cultures of *Thalassiosira pseudonana*; but when tested in mixed cultures with *Dunaliella tertiolecta*, the competitive success of *T. pseudonana* is decreased (Fisher et al. 1974, Mosser et al. 1972). Experiments using various concentrations of cadmium illustrate similar consequences of competition between plant species. In the presence of 0.05 mg/kg cadmium, the aquatic weed *Salvinia* does not survive. But when grown in competition with *Lemna*, *Salvinia* survives at this level of cadmium (Hutchinson and Czyrska 1975).

In addition, the confinement of test organisms is likely to create other stresses that will affect responses to the test substance. For example, although certain levels of the test chemical might be tolerated in the natural habitat, decreased mobility may affect the organism's response in laboratory tests. Therefore, even though results obtained in these test systems provide useful information about a species, the information must be applied with caution when used to formulate conclusions about organisms in their natural habitat. Even more care is required when conclusions are drawn about responses of other types of organisms by extrapolation from these test results. Many examples can be found in the literature of different responses to a chemical by different species, such as changes in diversity for an ecosystem (Bowes 1972, Hollister and Walsh 1973) and different effects on reproduction, life stages, and accumulation of pollutants (NRC 1979).

Indirect effects (e.g., effects on populations that depend on interactions with other species for survival, such as plant and herbivore relationships) cannot be detected by single-species tests. For example, a substance may not directly affect a test species, being only transformed by individual organisms. The transformation product might affect the interaction of this species with other components in the system (e.g., species A may become palatable to species B). The effect of 2,4-D on ragwort illustrates this point. Sublethal doses of this compound can increase sugar levels within the plant (Blodgett 1975) and render a naturally toxic plant more attractive to grazing cattle. Other indirect effects on plant communities have been studied also. In the presence of pollutants, increased incidences of insect attacks on trees have been reported (Stark et al. 1968) as well as increased incidences of disease (Treshow 1975) and parasitism (Heagle 1973).

An example of predator-prey interaction illustrates the importance of anticipating indirect effects (Taub 1976). If a chemical introduced into the natural environment reduces the growth of a prey population, so that both birth and death rates are effectively lowered by the establishment of an older age structure, the relative population size could remain the same, but the flow of biomass available in the system might be reduced. A predator population that relies on a particular level of flow of biomass could lose a substantial source of food. If that source were critical to the survival of the predator (i.e., no other sources were available), the population could become extinct, although no change is observed in the size of the prey population. Single-species tests that used either a predator or a prey would not detect this potential outcome.

Even in cases where laboratory tests on a single species clearly indicate a direct effect of a substance, this result could be masked in the natural habitat by concomitant effects on other components of the ecosystem. For example, in the laboratory a chemical may adversely affect the test species, A, resulting in increased mortality. In an ecosystem, however, the chemical could affect a predator of A by inhibiting its reproduction. Although the size of A is reduced in the laboratory, the "relief" from predation in the natural environment may compensate for this effect, thus resulting in only a slight overall change (either increase or decrease) in its population size.

Single-species tests give little attention to natural adaptability of a species. In the natural habitat, characteristics of populations can change over time as a result of a selection process that produces adaptation to complex and conflicting environmental stresses. An ecosystem also can adapt by changes in dominant species as abundance of coexisting species are altered (Hunding and Lange 1978). These adaptations do not evolve in isolation but depend on many system-produced limiting factors, both biotic and abiotic.

In addition, the adaptive capability of a heterogeneous, natural population may be quite different from that of a more homogeneous, standard laboratory test population. The reservoir of potential genetic changes found in wild populations may lead to adaptive changes, thus reducing the impact of the substance on the ecosystem. A few individuals may succumb, but the majority of them may survive the insult. In some cases, however, a natural population may not respond to the presence of a chemical at all. Evidence supports this possibility: data from laboratory tests with the standard set of aquatic invertebrates suggest that mirex is highly toxic to these species; however field data failed to corroborate these findings (NRC 1978).

The limited observations that are made in standard single-species tests have not always alerted scientists to potentially severe ecosystem consequences. Although results of lethal concentration (LC_{50}) tests for aquatic invertebrate species indicated low toxicity for polychlorinated biphenyls, subsequent field work and multi-species tests revealed a decrease in the diversity of invertebrate populations (Roberts et al. 1978).

Accumulation of DDT in the environment led to a surprising impact on raptors and fish-consuming birds. Adults could tolerate high concentrations of the chemical in the diet, but the compound interfered with eggshell formation, thus reducing the reproductive success of particular species. Low-level concentrations of DDT also can produce a variety of effects that are not *usually* observed in standard single-species tests. There is evidence that DDT alters temperature preferences in some fish (Anderson 1971); without specific tests for behavioral changes this type of effect will remain undetected. In a fish hatchery, mature lake trout can appear normal in all respects despite the presence of DDT residue in tissue samples; eggs hatch normally, but the fry die just before feeding commences, as a result of increased internal absorption of yolk nutrients contaminated with DDT (Burdick et al. 1964).

Laboratory test results also can appear more severe than effects observed in field studies. Subchronic tests using several aquatic invertebrate species exposed to methoxychlor demonstrated that some of these species were affected adversely at concentrations of 0.2 µg/l (Eisele 1974). A one-year field study investigating exposure in streams yielded additional information on population effects and interactive responses.

Only very subtle changes were detected in individual species at 0.2 μg/l in the stream environment, and multi-species interactions, such as predator-prey relationships, appeared unaffected (Eisele and Hartung 1976).

Pollutant realism, which also is important to assessment strategy, is achieved when all potential forms of the compound are considered for testing in the laboratory. In addition, a chemical rarely is present as the sole contaminant. The presence of other substances can produce synergistic or antagonistic effects that should be carefully considered.

Ecosystem processes can affect the chemical as well as its impact; its effective concentration or direction of movement can be changed before it reaches a target species. The system influences the effective concentration of a substance by partitioning it into any of several compartments. Depending on specific chemical and physical properties of the compound, storage in subsurface soil layers or in deep sediment may reduce its accessibility to a number of species. In aquatic systems dilution can occur after entry into the system, thus reducing the concentration that eventually reaches a species. Species that are not affected by a substance can accumulate it, removing the chemical from interaction with more sensitive components. Mobile species can remove the compound to extreme boundaries of the ecosystem and, depending on the density of organisms at this site, the impact may or may not be reduced.

Other interactions within an ecosystem also can modify the form of a pollutant. Physical properties of the system can cause changes in the physical state of the pollutant resulting either in its transportation out of the system without evident impact (e.g., volatilization), or resulting in different effects for apparently similar ecosystems. For example, the effects of DDT and dieldrin on natural populations of phytoplankton varied depending on whether the tests were done in Lake Erie or Lake Ontario (Glooschenko 1971). Biological processes can change the chemical composition of compounds, transforming them into products that may be less, or even more, toxic than the parent chemical.

Single-species tests, as they are now conducted, cannot provide the level of realism that is needed to assess adequately chemical effects on populations or multiple components of an ecosystem (Schneider 1980). Although it is possible to extrapolate some results obtained from one species to another and from one form of the chemical to another, the realism of interactive forces is not duplicated. At the highest level of complexity for population interactions, one finds such phenomena as the successional development of ecosystems. Single-species tests are obviously inappropriate to study effects of chemicals on this type of interaction. Similarly, such basic ecosystem properties as nutrient cycling, energy flow, and mineralization rates cannot be studied by tests that are not specific for those phenomena.

Ecotoxicology is a relatively recent subject of ecological research. It is most similar in scope and objectives to the field of radiation ecology, as both are concerned with the transfer, transformation, and effects of contaminants (radioactive or chemical) on ecosystems. The work of Chappel (1963), Ragsdale (1980), and Ragsdale and Rhoads (1974) illustrates that laboratory tests and models are insufficient to predict fates or effects of chemical substances. The ultimate fate or impact of a material is subject to variations in seasons, sites, and interactions of species (McCormick 1963, McCormick and McJunkin 1965, Platt 1965). Field and microcosm studies have demonstrated the influence of increasing ecological complexity and associated feedback effects upon the movement of chemical substances (Patten and Witkamp 1967, Ragsdale et al. 1968, Witkamp and Frank 1967, Witkamp and Merchant 1971).

SUMMARY

In general, ecosystems are complex, interlocking sets of components and processes, with properties that arise not just from the components themselves, but also from specific interactions among them. Therefore, it is not possible to characterize the response of any system to general or specific perturbations solely from the knowledge of the response of a few component parts. Vulnerability of ecosystems to the presence of chemicals depends on many factors, including chemical, physical, and biological properties of an ecosystem, and the characteristics and mode of entry of the chemical.

Single-species tests have several limitations that impair scientifically sound assessments of chemical impacts on ecosystems. The limitations include:

(1)   current laboratory tests examine only the responses of individuals, which are averaged to give a mean response for the test species;

(2)   with given constraints of limited finances and number of personnel, it is not possible to identify the most sensitive species or group of species;

(3)   the data are too limited in scope for extrapolations to be made for responses of other (even closely related) species;

(4)   indirect effects resulting from population or species interactions cannot be observed; and

(5)   conditions within which single-species tests are performed lack the realism of natural habitats.

Assessments of the potential toxicity of a chemical for particular ecosystems are extremely complex. Single-species tests, if appropriately conducted, are useful in evaluating a limited but valuable number of phenomena affecting an ecosystem. However, they must be combined with tests that can provide data on population interactions and ecosystem processes.

REFERENCES

Anderson, J.M. (1971) Assessment of the effects of pollutants on physiology and behavior. Pages 307-320, Proceedings of the Royal Society of London B177.

Batterton, J.C., G.M. Boush, and F. Matsumura (1972) DDT inhibition of NaCl tolerance by bluegreen alga, *Anacystis nidulans*. Science 176(4039):1141-1143.

Blanck, H.G. Dave, and K. Gustafsson (1978) An Annotated Literature Survey of Methods for Determination of Effects and Fate of Pollutants in Aquatic Environments. Report from the National Swedish Environment Protection Board.

Blodgett, J.E. (1975) Ecosystem effects of environmentally dispersed pollutants. *In* Effects of Chronic Exposure to Low Level Pollutants in the Environment. Committee on Science and Technology, U.S. House of Representatives, 94th Congress. Washington, D.C.: U.S. Government Printing Office.

Bowes, G.W. (1972) Uptake and metabolism of 2,2-bis-(p-chlorophenyl)-1,1,1,-trichloroethane (DDT) by marine phytoplankton and its effect on growth and electron transport. Plant Physiol. 49:172-176.

Burdick, G.E., E.J. Harris, H.J. Dean, T.M. Walker, J. Skea, and D. Colby (1964) The accumulation of DDT in lake trout and the effect on reproduction. Trans. Am. Fish. Soc. 93(2):127-136.

Chappel, H.G. (1963) The effect of ionizing radiation on *Smilax* with special reference to the protection afforded by their production of underground vegetative structures. Pages 289-294, Radioecology, edited by V. Schultz and A.W. Klement, Jr. New York: Reinhold Publishing Company.

Eisele, P.J. (1974) Effects of Methoxychlor on Stream Invertebrate Populations and Communities. Ph.D. Thesis, The University of Michigan, Ann Arbor, Michigan. (Unpublished)

Eisele, P.J. and R. Hartung (1976) The effects of methoxychlor on riffle invertebrate populations and communities. Trans. Am. Fish. Soc. 105(5):628-633.

Fisher, N.S., E.J. Carpenter, C.C. Remsen, and C.F. Wurster (1974) Effects of PCB on the interspecific competition in natural and gnotobiotic phytoplankton communities in continuous and batch cultures. Microbiol. Ecol. 1:39-50.

Glooschenko, W.A. (1971) The effect of DDT and dieldrin upon ^{14}C uptake by in situ phytoplankton in lakes Erie and Ontario. Pages 219-233, Proceedings of the 14th Conference of Great Lakes Research. Burlington, Ontario, Canada: International Association Great Lakes Research.

Heagle, A.S. (1973) Interactions between air pollutants and plant parasites. Rev. Phytopathol. 11:365-388.

Hollister, T.A., and G.E. Walsh (1973) Differential responses of marine phytoplankton to herbicides: Oxygen evolution. Bull. Environ. Contam. Toxicol. 9:291-295.

Hunding, C. and R. Lange (1978) Ecotoxicology of aquatic plant communities. Pages 239-255, Principles of Ecotoxicology, edited by G.C. Butler. Chichester, England: John Wiley & Sons.

Hutchinson, T.C. and H. Czyrska (1975) Heavy metal toxicity and synergism to floating aquatic weeds. Verh. Internat. Verein Theor. Angew. Limnol. 19:2102-2111.

McCormick, J.F. (1963) Changes in herbaceous community during a three-year period following exposure to ionizing radiation gradients. Pages 271-276, Radioecology, edited by V. Schultz and A.W. Klement, Jr. New York: Reinhold Publishing Company.

McCormick, J.F. and R.E. McJunkin (1965) Interaction of gamma radiation and other environmental stresses upon pine seeds and seedlings. Health Phys. 11:1643-1652.

Mosser, J.L., N.S. Fisher, and C.F. Wurster (1972) Polychlorinated biphenyls and DDT alter species composition in mixed cultures of algae. Science 176:533-535.

National Research Council (1978) Kepone/Mirex/Hexachlorocyclopentadiene: An Environmental Assessment. Committee for Scientific and Technical Assessments of Environmental Pollutants, Environmental Studies Board, Commission on Natural Resources. Washington, D.C.: National Academy of Sciences.

National Research Council (1979) Polychlorinated Biphenyls. Appendix D. Environmental Studies Board, Commission on Natural Resources. Washington, D.C.: National Academy of Sciences.

Patten, B.C. and M. Witkamp (1967) System analysis of cesium-137 in terrestrial microcosms. Ecology 48:813-824.

Platt, R.B. (1965) Radiation effects on plant populations and communities: Research status and potential. Health Phys. 11:1601-1606.

Ragsdale, H.L. (1980) The utility of single species and ecosystem tests in assessing the environmental impact of radionuclides. Commissioned paper prepared for the Committee to Review Methods for Ecotoxicology, Environmental Studies Board, Commission on Natural Resources, National Research Council. (Unpublished)

Ragsdale, H.L. and W.A. Rhoads (1974) Four-year post exposure assay of vegetation surrounding Project Pinstripe: Demonstration of the utility of delayed damage appraisals. Rad. Bot. 4:229-236.

Ragsdale, H.L., J.P. Witherspoon, and D.J. Nelson (1968) The Effects of Biotic Complexity and Fast Neutron Radiation on Cesium-137 and Cobalt-60 Kinetics in Aquatic Microcosms. Report No. ORNL-4318. Oak Ridge, Tenn.: Oak Ridge National Laboratories.

Roberts, J.R., D.W. Rodgers, J.R. Bailey, and M.A. Rorke (1978) Polychlorinated Biphenyls: Biological Criteria for an Assessment of Their Effects on Environmental Quality. Associate Committee on Scientific Criteria for Environmental Quality. Ottawa, Canada: National Research Council of Canada.

Schneider, R. (1980) Classes of ecotoxicological tests: Their advantages and disadvantages for regulation. *In* Working Papers for the Committee to Review Methods for Ecotoxicology. Available in limited supply from the Environmental Studies Board, Commission on Natural Resources. Washington, D.C.: National Academy Press.

Stark, R.W., P.R. Miller, F.W. Cobb, Jr., D.L. Wood, and J.R. Parmeter, Jr. (1968) Photochemical oxidant injury and bark beetle (Coleoptera: Scolytoidea) infestation of Ponderosa pine. I. Incidence of bark beetle infestation in injured trees. Hilgardia 39:121-126.

Taub, F.B. (1976) Demonstration of pollution effects in aquatic microcosms. Intern. J. Environ. Studies 10:23-33.

Treshow, M. (1975) Interactions of air pollutants and plant disease. Responses of Plants to Air Pollutants, edited by J.B. Mudd and T. T. Kozlowski. New York, N.Y.: Academic Press.

Witkamp, M. and M.L. Frank (1967) Cesium-137 kinetics in terrestrial microcosms. Pages 635-643, Proceedings of the Second National Symposium of Radioecology, edited by

D.J. Nelson and F.C. Evans, Jr. Springfield, Va.: National Technical Information Service.

Witkamp, M. and V.A. Merchant (1971) Effects of light, temperature and soil fertility on distribution of manganese-54 and cesium-137 in producer-consumer microcosms. Pages 204-208, Radionuclides in Ecosystems, edited by D.J. Nelson. Springfield, Va.: National Technical Information Service.

2 Factors Influencing the Fate of Chemicals

Precise predictions of the effects of chemicals on ecosystems depend in part on careful consideration of the fate of the chemical. Test conditions using single- or multi-species systems should include the concept of pollutant realism as discussed in Chapter 1. This chapter reviews some of the major factors that influence chemical fates and emphasizes their importance in evaluations of potential hazards.

CHEMICAL AND PHYSICAL FACTORS

The movement of chemicals into an ecosystem often is envisioned as a simple, direct, physical transfer from some activity related to humans. This, however, is only the first step of a complex process. If the chemical and physical properties that influence movement of a chemical within an ecosystem are well understood, the concentration of that chemical at any point in the system can be estimated.

Movement of chemicals into and between various media such as water, biota, and the atmosphere is a continuous process. Major routes for input to the atmosphere include (1) direct emission from such sources as manufacturing or processing plants (as dust, smoke, or vapor), fires, drift of sprays, vehicle exhaust; and (2) volatilization from surfaces of biota, soil, rocks, or structures made by humans. Once airborne, a chemical is transferred to terrestrial or aquatic sites through wet and dry deposition. Compounds also can be directly transported to terrestrial sites via land application of a chemical, accidental spills, waste disposal, landfill operations, and transport by organisms. Major entrance routes to aquatic systems include (1) direct input through intentional applications, accidental spills, discharge from industrial sources, and sewage disposal sites, (2) runoff from terrestrial sites, and (3) fallout from the atmosphere.

The fate of a substance is influenced by its own chemical and physical properties as well as by those of the ecosystem it enters. Important factors include: water solubility; vapor pressure; rates of volatilization, hydrolysis, photolysis, and sorption-desorption; octanol/water partition coefficients; boiling and melting points; ambient temperature, moisture, and humidity levels; wind velocity; biodegradation rates; conjugation rates; and leaching and dissipation characteristics (Miller 1978). The initial distribution of the substance in the environment is of considerable importance as well and must be given as much attention as the processes of transfer and transformation described here.

ENVIRONMENTAL PARTITIONING

To understand how a chemical is distributed within an ecosystem we must understand its affinity to various components of the system. With such understanding, transformation and degradation rates for a compound can be integrated with fugacity equations to provide information regarding the expected concentrations in various components of the ecosystem (MacKay 1979).

The movement of a chemical within an ecosystem and the tendency of that chemical to accumulate at specific sites (e.g., in sediment or at particular levels of a food chain) can be closely related to basic hydrophilic characteristics. If a compound is highly hydrophilic, it is less likely to move from water to other environmental media, such as soil, air, or biota. The potential for movement between media can be measured using easily calculated partition coefficients. A short discussion of them follows; a more detailed treatment is found in Appendix A, taken from McCall et al. (1980a).

Soil Sorption Constant (K_{oc})

Movement of chemicals through soil is a complex phenomenon that is influenced by a number of factors, including patterns of rainfall; rates of evaporation; properties of soil, such as surface area, amounts of clay and organic matter, pH, and temperature; and properties of the chemical, such as pK values, and hydrophilic and molecular characteristics. Despite this complexity, partition coefficients can be calculated and can serve as tools for identifying points of impact and potential movement of chemicals through soils.

The sorption constant (K_{oc}) relates the amount of chemical in soil to the amount in water. It can be used to rank chemicals according to their potential for leaching from soil (McCall et al. 1979). The sorption constant is based on the total organic carbon present in the soil and thus can be determined with a knowledge of carbon content independent of soil type. In modeling the movement of a chemical through soil, a sorption coefficient value (Kd) can be calculated for any given organic carbon content using K_{oc} (McCall et al. 1980b). Evidence indicates that the distance a compound moves through a soil column is inversely proportional to the sorption coefficient. (See, for example, McCall et al. 1980b.)

Reciprocal of Henry's Law Constant (K_w)

Henry's Law constant (H) represents the ratio of the concentration of a chemical in air to that in water and describes the distribution of a chemical under equilibrium conditions. The reciprocal of this constant (K_w) is related to transfer rates (from water to air) and is associated with the loss of chemicals from water due to volatilization. Because chemicals have low solubilities in water and thus low K_w values, those compounds with low vapor pressures can partition into air to a significant extent. Knowing K_w can alert investigators to the potential for volatilization of a chemical from aqueous systems.

Bioconcentration Factor (BCF)

The concentration or accumulation of chemicals within biota should be considered when determining environmental partitioning. The extent of biotic partitioning depends on the hydrophilic, lipophilic, and organophilic characteristics of the compounds under consideration. Various measurements or estimates of *BCF* can be made.

Aquatic systems permit the most accurate measurement of *BCF*, defined here as the ratio of the concentration of a chemical in aquatic organisms (μg chemical/g organism) to the amount in water at equilibrium (μg chemical/g water). Different species concentrate particular chemicals to differing degrees, but the relative ranking of any group or class of chemicals (e.g., high to low biotic concentration) tends to remain constant for all species.

For lipophilic compounds, *BCF* often can be estimated using a n-octanol/water partition coefficient (Neely et al. 1974, Veith et al. 1979). This coefficient can be determined experimentally or estimated based on one of two methods: use of water solubility values (Chiou et al. 1977) or an analysis of chemical structure (Fujita et al. 1964, Leo et al. 1971). Because these methods only estimate *BCF* values, they must be used with caution until more direct measures are developed.

For chemicals that are concentrated within organisms by mechanisms of active transport or transfer through food chains, *BCF* values can be determined only by direct observation under experimental conditions.

Use of Partition Constants

Partition constants can be combined in equilibrium models to estimate the distribution of a chemical in any ecosystem. Correlations between partition constants and other measurements of the hydrophilicity of chemicals have been made and verified (Briggs 1973, Chiou et al. 1977, Chiou et al. 1979, Fujita et al. 1964, Hamelink et al. 1971, Karickoff et al. 1979, Kenega and Goring 1978, Leo et al. 1971, Neely et al. 1974, Tulp and Hutzinger 1978). Using such relationships it is possible to estimate some aspects of the distribution of a chemical (Branson 1978, p. 59).

Most distribution models assume that equilibrium is reached in all compartments. This is not always true in nature; transfer rates between system components may be slower than rates of transformation or degradation within components. Much insight, however, can be gained when partition coefficients are combined in models and the net effect is evaluated. For example, a chemical may have a low K_w value, suggesting a tendency for volatilization. If it also has a high K_{oc} value, suggesting high sorption in aquatic systems, this factor may decrease the volatility potential of the chemical. Combinations of such partition parameters can determine the potential distribution and accumulation sites within an ecosystem. If these sites are known, appropriate tests can be selected for evaluating potential effects of a chemical.

Swain (1980) has studied the extensive environmental partitioning of polychlorinated biphenyls in the Great Lakes. This case study illustrates the usefulness of knowing both properties of separate system components and the interrelationships between them, and serves as a guide for developing an understanding of the fate, transfer, and accumulation of compounds within ecosystems.

TRANSFORMATION OF CHEMICALS

In addition to information on the distribution patterns of a chemical, it is important to identify transformation rates within compartments (e.g., soil, air, water, biota) of an ecosystem. Because transfer rates between components are generally ignored when calculating rates of transformation, care must be taken in applying estimates of those rates to any model of chemical distributions within an ecosystem. Rate constants can be estimated for all types of transformation processes: hydrolysis, oxidation, photolysis, and microbial degradation. The rate at which the chemical disappears from the system, however, will be a function of a diverse array of transfer and transformation processes within and among several compartments. The fate of a chemical, therefore, cannot be documented merely by examining single compartments without understanding the transfer processes across compartments.

Precise estimates of transformation and degradation rates occurring in nature are difficult to make because a multitude of pathways are involved. These rates are influenced by the physical, chemical, and biological characteristics of the ecosystem, and although individual reactions may obey first-order kinetics, aggregate reactions in a system often do not.

PHOTOLYSIS

Photolytic reactions can occur in the atmosphere, on vegetative and nonvegetative surfaces, and in various depths of water. The photochemical process occurs in three stages. The first is the absorption process: the chemical absorbs energy in the ultraviolet-visible spectrum of light, producing an excited molecule. The second stage is the primary photochemical process; if the excited molecule does not return to its original energy level, it undergoes a chemical reaction that could include fragmentation (e.g., formation of free radicals), rearrangement, or ionization. In the third stage, the active form of the chemical (e.g., free radicals) reacts with other molecules in the medium such as oxygen and water. Photolysis also may occur through sensitized reactions, in which those molecules that are efficient absorbers of light energy act as catalysts, transferring energy to less sensitive molecules. These are then degraded through absorption of the transferred energy. A review of photochemical transformations, including examples of phototransformation of selected chlorinated compounds, can be found in Korte (1978).

MICROBIAL DEGRADATION

Activities of microorganisms can alter the toxicity of chemicals, either through mineralization or cometabolic processes. Mineralization refers to the conversion of organic compounds to inorganic products by microbial metabolic processes. These reactions generally result in a less toxic material; however, there are some toxic substances or chemicals of ecological concern, such as sulfide and nitrate that may also be produced. Growth of heterotrophic microorganisms is associated with mineralization, because the process provides the organisms with an energy source and carbon molecules for growth. As the number and biomass of microorganisms increase, the chemical disappears at a faster rate; thus, the rate of mineralization tends to parallel the rate of microbial growth. Moreover, once the chemical has induced growth of the microbial population, subsequent inputs of the compound are mineralized more readily. Most natural products are degraded by this process, some very rapidly (e.g., simple sugars and certain aromatic compounds) and others more slowly (e.g., polyaromatics).

Cometabolism refers to the breakdown of compounds that are not used by microorganisms as sources of energy, carbon, or some other essential nutrient. Because the microorganism derives no benefit from the transformation, population size and biomass do not increase and chemicals subject to this process are not destroyed at increasing rates with time. Moreover, in contrast to mineralization, the rate of degradation does not increase after initial input of the chemical. Cometabolized products often accumulate, sometimes serving as substrates for other organisms in the ecosystem. Recent evidence suggests that a variety of synthetic compounds can be subject to cometabolic processes (Alexander 1979).

The evidence that cometabolism is responsible for transformation of many synthetic compounds is obtained from observations that (1) the compounds are transformed in nonsterile but not in sterile soils or water indicating that the process is biological; (2) no organisms obtained from the test medium can use the compound as a source of energy, carbon, or some other essential nutrient; (3) the compounds are not degraded at increasing rates with time; and (4) the rate of cometabolism does not depend on prior additions of the compound to the medium. Using these lines of evidence, compounds such as DDT, aldrin, 2,4,5-T, and PCBs appear to be transformed by cometabolic processes (Alexander 1979).

OTHER TRANSFORMATIONS

Activation refers to the conversion of a compound of low or no toxicity to a product that is highly toxic. The process may be abiotic (as in the activation of dimethylamine to dimethylnitrosamine), or it may result from microbial activity. Few generalizations can be made regarding the types of chemicals that may be subject to activation because little attention has been given to this transformation process.

Transformation of chemicals can change the nature of their toxicity. The transformation can lead to the formation of chemical species that affect groups of organisms different from those that are affected by the original substance. The conversion of pentachlorobenzyl alcohol into chlorinated benzoic acid illustrates this phenomenon; the parent molecule is antifungal in nature, but the chlorinated benzoic acid is phytotoxic (Ishida 1972). Indeed, these acids are marketed as herbicides. Similarly, in soil the fungicide thiram is degraded to yield an intermediate product that is then nitrosated to produce the carcinogen nitrosamine (Ayanaba et al. 1973).

Other biotic transformation processes include polymerization, conjugation, oligomerization, and dimerization. Microorganisms frequently polymerize simple aromatic compounds into more complex polyaromatics (Martin and Haider 1976). Many simple aromatics are initially toxic or might be modified to yield toxic products, but the behavior of the polyaromatics is unknown. Conjugation, or addition reactions, are illustrated by the nitrosation process and in acylations and alkylations. Several pesticides are known to be converted to formyl or acetyl derivatives, presumably as a result of biological activities in soil (M. Alexander, Cornell University, Ithaca, N.Y., personal communication, 1980). Aklylation of chemicals is largely restricted to their methylation, e.g., the methylations of sulfur (Bremner and Steele 1978), selenium (Doran and Alexander 1976), and arsenic (Cheng and Focht 1979). A number of synthetic chemicals are transformed by oligomerization reactions (Bollag et al. 1977). Dimerization is a common transformation of amino compounds in soil (Bartha and Pramer 1970).

In vertebrates, oxidative reactions followed by conjugations to glucuronides and organic sulfates predominate. The mixed-function oxidase system is particularly important in the initial oxidation of aromatic compounds, and these metabolic processes may either decrease or increase toxicity. Reductive pathways and many other types of conjugation also have been reported, but they rarely occur in vertebrates (Williams 1960).

Clearly, many transformation products can be generated, and their rates of formation or destruction are affected by biotic and abiotic properties of the environment. Therefore it is important to consider particular environmental characteristics as well as the identity of transformation products to determine accurately the potential toxicity of any substance. In addition to tests of the parent chemical, one must consider the effects of the transformation products as well.

FATE CONSIDERATIONS FOR TERRESTRIAL ECOSYSTEMS

Both vegetative and nonvegetative surfaces are important receptors of chemicals deposited in terrestrial ecosystems. Because vegetation is the primary supply of food and fiber for human beings and other animals, it has an important role in transferring chemicals within a system. Both wet and dry depositions are primary input routes. Once on the vegetation, chemicals may follow several pathways: (1) volatilization back into the atmosphere, (2) washing off to soil or other surfaces, (3) adsorption to plant surfaces, and (4) absorption and translocation into plants. Chemicals may be degraded by biological processes within plants or by photochemical processes while the substance is on the surface of plant material. As noted earlier the metabolite or transformation product also may be toxic. If the chemical remains intact in or on the plant, its fate depends, in part, on the nature of the plant; a chemical that remains in the plant may produce an effect that changes the attractiveness of the plant to other components of the ecosystem. There are other ways that the fate of chemicals is determined. For example, when the vegetation is cycled (drops foliage, dies, is burned, or is consumed by herbivores), the material will enter another environmental pathway. A chemical also can be translocated to roots of vegetation where, over long periods of time, it could be released to the soil and made available to other parts of the system. Bioconcentration in certain parts of the plant could occur and present potentially toxic levels of a chemical to herbivores.

The nonvegetative surfaces of terrestrial ecosystems can receive large quantities of chemicals through atmospheric deposition, sprays, spills, waste disposal, and landfill operations. Once chemicals reach these sites the material might be (1) transported away from the point of application by wind, water, or movement of organisms, (2) volatilized, (3) leached through soil, (4) adsorbed onto or absorbed into soil, or (5) taken up by plants. The availability of chemicals remaining in terrestrial systems to various organisms then becomes important. Factors that control this availability include the chemical nature of each compound, the nature of the adsorbing surface (e.g., amount and type of organic matter and clay), the nature and location of the site (e.g., type, abundance, and susceptibility of biota), and the amount of material involved; small quantities of material may be rendered innocuous while large amounts may exceed the detoxifying or adsorbing capacity of the site.

Soils also play an important role in the movement of chemicals in terrestrial systems (Korte 1978). Transport phenomena within soil are strongly influenced by accumulation of detergents and inorganic salts and by such factors as type of soil, moisture content, and, to some extent, weather conditions.

FATE CONSIDERATIONS FOR AQUATIC ECOSYSTEMS

A chemical in water can exist in one or more of four states: (1) in solution, (2) adsorbed to a biotic or an abiotic surface, (3) suspended in the water column, or (4) incorporated (and perhaps accumulated) into living or dead organisms. Compounds in these states may be adsorbed (primarily to sediment), accumulated, diluted (most rapidly in streams or from single inputs), degraded (chemically, microbially, or photolytically), transported, or volatilized.

Hydrophobic chemicals tend to be enriched in the surface micro-layer or adsorbed onto suspended colloids, micro-particulates, and sediments. These compounds may be irreversibly adsorbed to surfaces (thus unavailable for further transformation); in equilibrium between water, other adsorbing surfaces, and living organisms; transported; degraded in place; or covered with other sediment. Compounds covered by sediment can be lost to the system until movement of organisms disrupts the sediment cover, permitting redistribution of the chemical. Compounds that are bioaccumulated may be stored in various tissues, metabolized, released (e.g., excreted) back into water, transported through the food chain, or transported to other aquatic sites by mobile species. The desorption and dissolution of many elements from sediment is highly pH-dependent; thus small changes in pH can result in large changes in concentrations present in the water column.

VARIABLES AFFECTING FATES

Several sources of variability affect the fate and detection of a chemical. Of particular importance are variations in the rate of input and variability of exposure and concentration between test conditions and natural conditions.

RATES OF INPUT

The physical input of chemicals into ecosystems can be one of the most variable characteristics and one of the most difficult to estimate. The difficulty arises not in quantifying the actual initial input, but from variations in distribution among and within systems. For example, accurate calculations can be made for the amount of pesticides applied annually to agricultural, pastoral, and forest systems. Less accuracy is possible in determining the percentage of total application that leaves these systems, for example through volatilization or water run-off. Similarly, fairly accurate estimates can be made for chemicals that are airborne or discharged directly into waterways. Once the chemical enters an ecosystem—terrestrial or aquatic—however, the rate and extent of dissipation are often unknown or, at best, poorly estimated.

Intermittent inputs, those varying over time, pose particular problems in determining the rate for a certain period. Examples of intermittent inputs include runoff from terrestrial sites during intense rainfall, accidental spills, and periodic discharges from manufacturing plants. The impacts of these events are subject to many variables, including differences in chemical form (e.g., pure or in combination with surfactants or other additives), differences in chemical characteristics (e.g., solubility, volatility, n-octanol/water partition coefficient), variable characteristics of the environment (e.g., adsorption capacity), the time between input and rainfall events, and concentration of chemical per unit area or volume. Of course, chemicals also may enter an ecosystem at a steady rate, as in industrial emissions to the atmosphere or water and discharges from sewage treatment facilities. While rates of constant input at any particular time are better known than rates of intermittent input, the rates and degrees of distribution of both types depend on the characteristics of the chemical, the receiving system, and interactions between them.

RATE OF EXPOSURE

Another source of variability when determining fates of chemicals arises during toxicological testing. A particular dose or concentration and expected duration of exposure must be related to the response observed in the test organism. Preferred protocols for laboratory tests should specify dynamic exposure conditions that maintain constant levels of a chemical in the test system. Chemical, biological, or financial constraints may prevent the use of this approach, thereby necessitating the development of static exposure systems (see Figure 2.1). In such systems a known concentration is added at the initiation of an experiment. Subsequent measurement of residue levels may or may not be taken. Concentrations of the chemical in these systems tend to decrease over time, simulating a single exponential decay function. In tests that involve more than one species, a complex change in chemical concentration and form also may result.

Concentration profiles derived from dynamic and static systems in laboratories have limited application to natural ecosystems because they represent simplified examples that rarely occur under natural conditions. Levels of chemicals in the natural environment can vary in a complex fashion (see Figure 2.2) as a result of many factors: variable inputs of the chemical; recycling through detritus webs, and among species in food chains; and changes in such environmental conditions as pH, temperature, and moisture.

The three concentration profiles (dynamic, static, and natural) are different and these differences complicate any attempt to relate responses of organisms that are tested using one profile to responses that might be expected from exposure to another. Approximate correlations are possible for limited sets of conditions, however. For example, results of static tests of short duration involving chemicals that are highly water soluble can be correlated with results from dynamic tests. But if rapid biodegradation or hydrolysis were to occur, or if the chemical were not soluble in water, the correlations would be poor. Divergence between results from the two tests also would be greater as exposure time is increased.

Comparisons of results from constant-level exposures and variable exposures are difficult. If variations in the latter are relatively small, then time-weighted averages of these concentrations may be used. This relationship is based on Haber's rule, which states that for small differences in concentration, the percentage of organisms affected (E) equals the product of a constant (K), exposure time (T), and concentration (C): $E = KCT$.

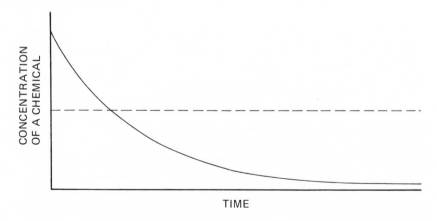

FIGURE 2.1   Time-concentration profile for dynamic (---) and static (-) exposures.

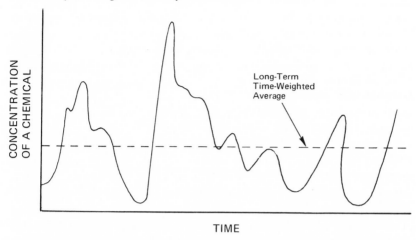

FIGURE 2.2 Time-concentration profile for natural conditions.

UNEQUAL CONCENTRATIONS

To allow the use of laboratory results, the assumption often is made that the concentration used in laboratory tests is identical to the same concentration measured in natural ecosystems. This is not always a valid assumption. For example, inorganic chemicals are invariably introduced into test systems in a dissolved form. Heavy metals usually are introduced as acid salts that have been kept in stock solution at very low pH. When these solutions are introduced into a test system, the pH increases because of the buffering capacity of the test water. As a consequence, the form of the heavy metal is likely to change and can be considerably less soluble than the original material. The new form is not necessarily stable, but can change gradually until it resembles the chemical that is detected in nature. The time required for these changes, however, may be much longer than either the residence time of the test water in the laboratory system or the duration of the test. Thus, the form in which a substance is dispersed can be quite different in laboratory systems from that present under ambient environmental conditions, even if total concentrations are identical.

Similar difficulties exist with organic chemicals that are not readily soluble in water. These compounds usually are introduced into test systems dissolved in relatively nontoxic organic solvents (e.g., acetone or ethanol). Occasionally surfactants, such as Triton X-100, are added. The assumption is made that organic chemicals added in this fashion remain in solution, but this is not necessarily valid. Again, the assumption that equal concentrations, as measured in laboratory systems and under field conditions, represent equal degrees of biological availability cannot be supported. The proportions of the test chemical found in free solution as micelles or adsorbed to colloids and other particulates are likely to vary between test and ambient conditions. Because the physical form of a compound influences both its biological availability and its fate, the interpretation of test results should be made with caution, and the use of systems that minimize formation of micelles should be encouraged.

Commercial formulations of most chemicals contain several intentionally or unintentionally added ingredients that may be present in varying concentrations. In some instances these additions could significantly influence the impact of the test substance. Rarely is the toxicity of these added components known, nor is it determined during the testing process. Furthermore, because the manufacturer or distributor may not provide information on the carrier or solvent used in manufacture of the commercial formulations, the effects of these chemicals are often neglected.

Test chemicals often contain contaminants that are similar in structure or were precursors used in synthesizing the parent compound. These contaminants may vary in identity and in quantity among manufacturers and within batches from a single manufacturer. Occasionally the contaminant is far more toxic than the parent material. This is evident in the occurrence of dioxins in commercial preparations of 2,4,5-T and in the high content of nitrosamines once present in certain herbicides.

For these reasons, evaluations of the toxicity of new classes of chemicals or individual compounds should be initiated only after a careful analysis of the chemical and physical properties of the molecule and of materials that may be associated with it. The experimental approach, test system, and means of applying the chemical should be designed carefully to obtain valid data and meaningful interpretations.

SUMMARY

Chemical and physical properties of both the substance and the environmental medium are important in identifying pathways of potential transport. Such relationships as distribution of a material between soil and water (K_{oc}), between air and water (K_w), within biological components (BCF), and the potential for formation of transformation products should be determined. Ecosystems exhibit special properties that also affect the fate of a chemical. These should be considered carefully in assessing the potential toxicity of substances. Variability in estimates or measurements occurs at many points, from the source of discharge of a substance to individual responses to the presence of a chemical. Because of this variability, test conditions should be carefully designed and data should be interpreted with caution.

The distribution and subsequent fate of chemicals determine the dose delivered to the many biotic components of an ecosystem. Because dose and duration of exposure are of prime importance for determining the magnitude of an effect, accurate estimates of each are needed when possible. The most promising avenue for obtaining this information early in the evaluation process is the application of fugacity or partitioning equations combined with rates of degradation, transformation, and transfer.

REFERENCES

Alexander, M. (1979) Role of cometabolism. Pages 67-75, Microbial Degradation of Pollutants in Marine Environments, edited by A.W. Bourquin and P.H. Pritcherd. Report No. EPA-600/9-79-012. Gulf Breeze, Fla.: U.S. Environmental Protection Agency.

Ayanaba, A., W. Verstraete, and M. Alexander (1973) Formation of dimethylnitrosamine, a carcinogen and mutagen, in soils treated with nitrogen compounds. Soil Sci. Soc. Am. Proc. 37:564-568.

Bartha, R. and D. Pramer (1970) Metabolism of acylanilide herbicides. Adv. Appl. Microbiol. 13:317-341.

Bollag, J.M., R.D. Sjoblad, and R.D. Minard (1977) Polymerization of phenolic intermediates of pesticides by a fungal enzyme. Experientia 33:1564-1566.

Branson, D.R. (1978) Predicting the fate of chemicals in the aquatic environment from laboratory data. Pages 55-70, Estimating the Hazard of Chemical Substances to Aquatic Life, edited by J. Cairns, Jr., K.L. Dickson, and A.W. Maki. ASTM Special Technical Publication 657. Philadelphia, Pa.: American Society for Testing and Materials.

Bremner, J.M. and C.G. Steele (1978) Role of microorganisms in the atmospheric sulfur cycle. Advan. Microb. Ecol. 2:155-201.

Briggs, G.G. (1973) A simple relationship between soil adsorption of organic chemicals and their octanol/water partition coefficients. Pages 83-86, Proceedings: 7th British Insecticide and Fungicide Conference, sponsored by British Protection Council, 3 Vols. Reproduced by Boots Co., Ltd., Nottingham, England.

Chiou, C.T., V.H. Freed, D.W. Schmedding, and R.L. Kohnert (1977) Partition coefficients and bioaccumulation of selected organic chemicals. Environ. Sci. Technol. 11(5):475-479.

Chiou, C.T., L.J. Peters, and V.H. Freed (1979) A physical concept of oil-water equilibria for nonionic organic compounds. Science 206:831-832.

Doran, J.W. and M. Alexander (1976) Microbial formation of volatile selenium compounds in soil. Soil. Sci. Soc. Am. J. 40:687-690.

Cheng, C.N. and D.D. Focht (1979) Production of arsine and methylarsines in soil and in culture. Appl. Environ. Microbiol. 38:494-498.

Fujita, T., J. Iwasha, and C.J. Hansch (1964) A new substituent constant, π, derived from partition coefficients. J. Am. Chem. Soc. 86:5175-5180.

Hamelink, J.L., R.C. Waybrant, and R.C. Ball (1971) A proposal: Exchange equilibria control the degree of chlorinated hydrocarbons and are biologically magnified in lentic environments. Trans. Amer. Fish. Soc. 100:207.

Ishida, M. (1972) Phytotoxic metabolites of pentachlorobenzyl alcohol. Pages 281-306, Environmental Toxicology of Pesticides, edited by F. Matsumura, G.M. Boush and T. Misato. New York, N.Y.: Academic Press.

Karickoff, S.W., D.S. Brown, and T.A. Scott (1979) Sorption of hydrophobic pollutants on natural sediments. Water Res. 13:241-248.

Kenaga, E.E. and C.A.I. Goring (1978) Relationship between water solubility, soil sorption, octanol-water partitioning and concentration of chemicals in biota. Spec. Tech. Pub. 707, Proceedings: Third ASTM Symposium on Aquatic Toxicology. New Orleans, Louisiana, Oct. 17-18, 1978. Philadelphia, Pa.: American Society for Testing and Materials.

Korte, F. (1978) Abiotic processes. *In* Principles of Ecotoxicology, edited by G.C. Butler. Chichester, England: John Wiley & Sons, Inc.

Leo, A., C.J. Hansch, and D. Elkins (1971) Partition coefficients and their uses. Chem. Reviews 71(6):525-616.

McCall, P.J., R.L. Swann, D.A. Laskowski, S.A. Vrona, S.M. Unger, and H.J. Dishburger (1979) Prediction of chemical mobility in soil from sorption coefficients. Proceedings: Fourth ASTM Symposium on Aquatic Toxicology. Chicago, Illinois, Oct. 16-17, 1979. Philadelphia, Pa.: American Society for Testing and Materials. (Due for release in 1981)

McCall, P.J., D.A. Laskowski, R.L. Swann, and H.J. Dishburger (1980a) Partitioning of chemicals in model ecosystems. Commissioned paper prepared for the Committee to Review Methods for Ecotoxicology, Environmental Studies Board, Commission on Natural Resources, National Research Council. (Unpublished)

McCall, P.J., R.L. Swann, D.A. Laskowski, S.M. Unger, S.A. Vrona, and H.J. Dishburger (1980b) Estimation of chemical mobility in soil from liquid chromatographic retention times. Bull. Environ. Contam. Toxicol. 24:190-195.

MacKay, D. (1979) Finding fugacity feasible. Environ. Technol. 13(10):1218-1223.

Martin, J.P. and K. Haider (1976) Decomposition of specifically carbon-14-labelled ferulic acid: Free and linked into model humic acid-type polymers. Soil Sci. Soc. Am. J. 40:377-380.

Miller, D.R. (1978) General considerations. Principles of Ecotoxicology, edited by G.C. Butler. Chichester, England: John Wiley & Sons, Inc.

Neely, W.B., D.R. Branson, and G.E. Blau (1974) Partition coefficient to measure bioconcentration potential of organic chemicals in fish. Environ. Sci. Technol. 8:1113-1115.

Swain, W.R. (1980) An ecosystem approach to the toxicology of residue forming xenobiotic organic substances in the Great Lakes. *In* Working Papers for the Committee to Review Methods for Ecotoxicology. Available in limited supply from the Environmental Studies Board, Commission on Natural Resources. Washington, D.C.: National Academy Press.

Tulp, M. Th. M. and O. Hutzinger (1978) Some thoughts on aqueous solubilities and partition coefficients of PCB, and the mathematical correlation between bioaccumulation and physico-chemical properties. Chemosphere 7(10):849.

Veith, G.D., N.M. Austin, and R.T. Morris (1979) A rapid method for estimating log *P* for organic chemicals. Water Res. 13:43-47.

Williams, R.T. (1960) Detoxification Mechanisms: The Metabolism and Detoxication of Drugs, Toxic Substances and Other Organic Compounds. 2nd edition. New York, N.Y.: John Wiley & Sons, Inc.

3 Relevant Properties and Processes

The effects of chemicals on ecosystems cannot be predicted solely on the basis of single-species tests. Although effects of substances on particular species may be predicted with some precision using data generated by such tests, responses of organisms in a natural setting can be very different. Furthermore, effects that might not be measured or detected in simple single-species tests could have significant but unpredictable impacts on ecosystems.

A better understanding of how given levels of contaminants influence the structure and functions of an ecosystem is needed to make an adequate determination of whether a substance is potentially harmful. When possible, test systems should be developed on the basis of the concept of ecological realism as discussed in Chapter 1. The purpose of this chapter is to indicate the type of information needed to determine effects of chemicals on ecosystems. A review is presented of those properties and processes that might be affected by environmental contaminants and those that can influence the magnitude of the impact. All are aspects of the biological integrity of populations and ecosystems; a change in any of them would be suggestive of a chemical impact that might warrant regulation.

PROPERTIES VULNERABLE TO EFFECTS OF CHEMICALS

Certain properties or characteristics of ecosystems and populations may be particularly vulnerable to the introduction of chemicals. This section discusses aspects of each that are important in evaluating the toxic potential of chemicals. No attempt has been made to screen these characteristics on the basis of currently available, economical methodology; a number of scientifically justifiable concepts have been included under the assumption that suitable methodology can be developed.

POPULATION PROPERTIES

Mortality, Fecundity, and Rate of Growth

The effects of many chemical substances on the mortality, fecundity, and rate of growth have been documented. Chemicals introduced into a system can cause direct mortality of both target and nontarget species. Also there is good evidence that chemicals can reduce fecundity without affecting mortality. The evidence includes DDE effects on mallard ducks and ring doves (Heath et al. 1969, Haegele and Hudson 1973), DDT effects on predators (Henny 1977), avian responses to pesticides (Henny 1972), and PCB impacts on a variety of species (Roberts et al. 1978). Certain compounds may have different sex-specific effects: effects of methyl mercuric chloride on brine shrimp and reproduction are a good example (Cunningham and Grosh 1978). The impairment of reproduction is a subtle effect of chemical contamination, but ultimately it will result in a major effect on the population.

Changes in fecundity, survivorship, and mortality may be the most sensitive measures for evaluating effects of chemicals on populations (see Hutchinson 1978 for a thorough discussion of these functions, Canton et al. 1975). While it is difficult to measure these properties in most natural populations, values can be obtained with some accuracy in laboratory populations (most often using various invertebrates). Figure 3.1 shows the response of two of these functions (survivorship and fecundity) to changes in temperature and abundance of food. In principle, methodologies can be developed to use all these as measures of the effect of chemicals on populations.

Although minor changes in rates of growth have resulted from exposure to a variety of chemicals (Fendley and Brisbin 1977), the impact of such changes on subsequent survival of the population is unknown. Reduced growth rates can delay sexual maturity or can increase the susceptibility of populations to disease or predation (Friend and Trainer 1970). Individuals of abnormal size or with abnormal patterns of behavior have been observed to be selected preferentially by predators (G.W. Salt, University of California, Davis, personal communication, 1980; Cooke 1971). Rates of growth can be estimated by observing age distribution or size distribution and stages in the life history. The latter are particularly useful for invertebrates characterized by instar progression.

An example of instar analysis are illustrated in Figure 3.2. An increase in egg production is verified by population increases in subsequent life stages. At each life stage lengthening of instar periods will indicate periods most sensitive to a chemical even if there are no major changes in total population size. Two other examples of instar analysis are shown in Figures 3.3 and 3.4. In the first figure, there is a limited period of reproduction of only a few weeks each year producing a cohort of limited age and instar composition at any one moment. The cohort slowly passes through all the stages of development, producing a group of mature animals once a year. The second example shows a more complicated situation, with several generations occurring during one summer growing season. Instar analysis permits one to follow the success or failure of each generation and offers the possibility of determining reasons for the success or failure. Thus the possibility exists for investigating the effect of a chemical on these organisms.

The size of a population varies with changes in birth, emigration, immigration, and mortality rates. Large changes in these factors after introduction of highly toxic substances are immediately evident. Long-term effects that only slightly increase the rates of change, for example, by increasing susceptibility to disease or predation are not as evident and could be masked entirely by normal fluctuations in system parameters. Further impacts of toxic substances also may result in excessive population growth resulting from reduced competition among species (e.g., when dealing with species with similar habitat requirements or similar nutrient sources) or in overpopulation related to reductions in abundance of predators or herbivores.

Changes in age distribution above normally observed fluctuations are difficult to detect if wide ranges occur often. Detection of chemical effects would be possible, however, if it drastically altered these normal fluctuations. Although complete loss of a species is evidence of impacts on an ecosystem, determining the cause-effect relationships may be difficult.

Because impacts on invertebrate life cycles may be more significant than certain physiological effects on individuals, the ability of a population to withstand or acclimate to the presence of a chemical may be determined by experimental modification of the life cycle (Cole 1979). Therefore the timing of input and the persistence of a chemical relative to sequencing and complexity of the life stages of key species are critical factors in any prediction of impact.

1_x : survivorship, the percentage of animals reaching the given age (Logarithmic scale).

m_x : fecundity, offspring per hour per female of the given age.

Food is given as millions of *Chlorella* cells per ml. supplied every 12 hours at 20°C. The potential population growth rate resulting from the 1_x and m_x is given as *r*. Note that there is an optimal food supply because of a slight antibiotic effect of *Chlorella*.

For more information about life tables, see the text.

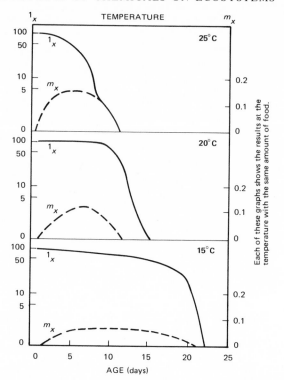

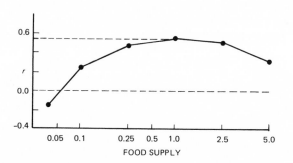

FIGURE 3.1   Population dynamics of the rotifer *Brachionus calyciflorus*.

SOURCE: Halbach (1974).

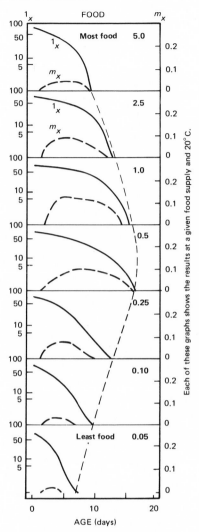

FIGURE 3.1 *Continued.*

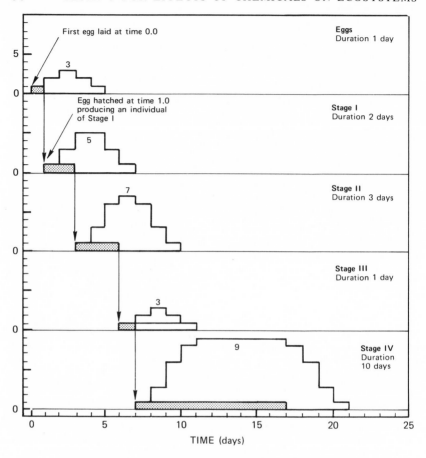

FIGURE 3.2   A hypothetical graphical model of instar analysis.

Phenotypic and Genotypic Variation

Short-term exposure to pollutants can produce a variety of phenotypic changes, both behavioral and physiological. Alteration in normal behavior patterns and in learning abilities have been observed following exposure of an organism to certain chemicals (Peterson 1977). Examples include the effects of methyl parathion (Farr 1978), mercury (Kania and O'Hara 1974), and mirex (Tagatz 1976) on predatory-prey interactions. Alteration in basal metabolic rates for individuals exposed to chemicals also has been reported; under stressful conditions the reduced rate can result in lowered potential for survival, altered behavioral response, and changes in energy availability during food-gathering activities. Sexual

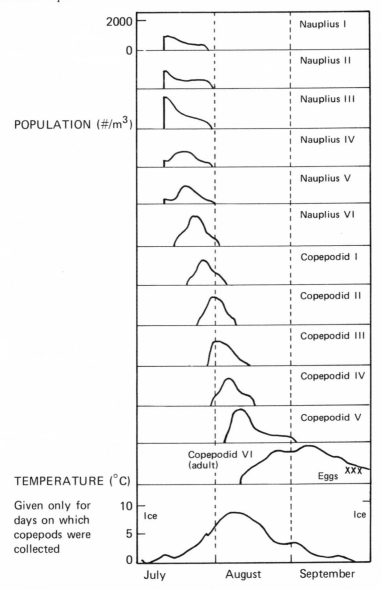

FIGURE 3.3 Life history of the calanoid copepod *Limnocalanus johnaseni* in Imikpuk, a shallow arctic lake near Point Barrow, Alaska.

SOURCE: Comita (1956).

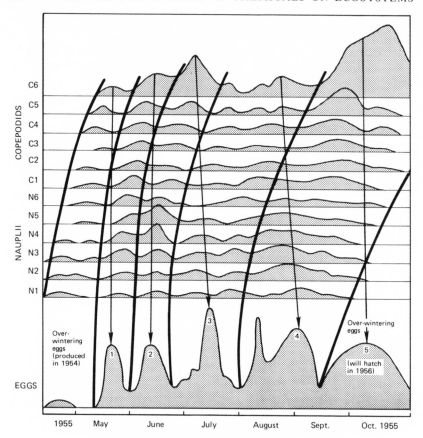

FIGURE 3.4   Seasonal life history of *Diaptomus siciloides*.

SOURCE: Data taken from Comita (1972).

maturity, fertility, and production of gametes in females can be affected by exposure to certain chemicals. (See, for example, Biesinger and Christensen 1972, Brungs 1969, Winner et al. 1977.)

For invertebrate species with short generation times, alteration in gene pools can occur rapidly. Selection for resistance to insecticides in nearly a hundred species is a well-documented example (NRC 1980, U.S. EPA 1975, Wagner 1974). If selection for resistance occurs as a response to chemical effects, then selection for other genetic characteristics also is possible. Species with short generation spans have an advantage in adapting to environmental contaminants.

The evidence for genetic selection among vertebrates is not as dramatic; in cotton-growing regions of the southern United States, for example, genetic changes have been observed in mosquito fish after their exposure to high levels of insecticides (Ferguson and Bingham 1966, Ferguson et al. 1966). When avian species are exposed to high doses of known environmental contaminants, genetic changes have been observed (Azevedo et al. 1972). Effects of pollutants on genetic factors of vertebrates, however, may not become evident until long after the chemicals have been degraded or dissipated throughout the environment.

Interactions of Stress Factors

Individual organisms have been shown to be more susceptible to diseases or other stresses following their exposure to environmental contaminants (Friend and Trainer 1970, Roberts et al. 1978, Treshow 1975). If enhanced susceptibility is evidenced by large changes in population size of species, then effects on the ecosystem also can be expected. Nutritional, physical, and social stresses affecting an organism influence its ability to resist the effects of a chemical. The impact of these stresses can be either synergistic, antagonistic, or additive. For example, enhancement of microsomal actions in liver tissue not only influences metabolic rates of ingested chemicals but also alters metabolic rates of natural steroids, thus affecting the physical condition of the organism.

Chemicals change normal functions of organisms; for example, a chemical could interfere with normal excretions of salt glands in waterfowl, thus affecting their ability to exist in highly saline environments (Friend et al. 1973). The salt tolerance capability of blue-green alga has been shown to be influenced by DDT (Batterton et al. 1972). Some chemicals could mimic the action of steroid hormones and thus disrupt hormonal balance of an organism.

Influence of Animal Behavior

As individual reproductive or courting behavior is altered, changes in population size can result; this, in turn, can alter the system. Complex releaser cues related to reproductive function in birds have been altered by chemicals (Haegele and Hudson 1973). Changes in behavior of fish may affect their survival or reproductive success (Sullivan et al. 1978). Chemicals can affect the ability of prey species to escape predators or the ability of the predator to detect the prey (Cooke 1971, Farr 1978, Goodyear 1972). If the prey is resistant to adverse effects of a substance but accumulates it in body tissue, the predator could be susceptible.

There is a distinct need to describe and define components that serve to integrate subtle effects of chemicals upon the behavior of individuals. The sum of these chemical effects on individuals could be evident at the level of populations and ultimately the ecosystem.

Migration

Initiation of migration by organisms is related to the age and physiological state of individuals, direct effects of temperature and light, density and aggressiveness of other animals in the local population, and availability of food. Local weather patterns sometimes are the primary factor initiating migration, but movement from the system often is a result of the need to avoid unsuitable local conditions, such as the presence of a chemical. For many species the cues leading to migration and the means of navigation during migration are not fully known. Cuing on chemicals, the sun, moon, and stars is documented and some organisms—such as fish, birds, and bees—have an ability to sense geomagnetic forces (Bullock 1973, Gould 1980, Keeton 1974, Lindauer and Martin 1968, Moore 1980, Presti and Pettigrew 1980, Southern 1974, Walcott 1974, Walcott et al. 1979, Wiltschko and Wiltschko 1972).

The potential impact of chemicals on migration occurs at the level of an individual through alterations in the physiological state or the ability to respond to important cues. Navigational capabilities of organisms also can be altered by toxic effects on neural pathways (Weiss 1959) and sensory receptors (e.g., Bardach et al. 1967).

Migration may be induced by large decreases in the size of producer populations and a subsequent reduction in energy flow. In these situations, secondary consumers migrate out of the system in search of suitable food and habitat, either on a temporary or permanent basis. For example, when large-scale mortality of a prey occurs, the migration of predators also can be expected.

Socialization during migration is important; if behavioral changes occur as a result of exposure to chemicals, migratory patterns could be influenced (Hansen 1969, Hasler and Wisby 1949, Jones et al. 1956, Lewis and Livingston 1977). Migration is a time of physical stress for many organisms. When long distances are covered and when there is a decrease in individual fitness prior to migration, survival could be jeopardized. Spring migrations to breeding grounds are also critical, as the physical condition of arriving females directly affects subsequent clutch size and reproductive success. If any physiological changes resulting from exposure to chemicals alter a species' ability to move during traditional times and periods between wintering and breeding sites, species that have adapted to these locations can be influenced greatly. Subtle cues related to sensitive receptors guiding migration may be disrupted by the presence of chemicals in the environment.

Diversity

The diversity of an ecosystem is defined as a measure of the variety of species in a system and takes into account the relative abundance of each species. Diversity can be measured for an entire food chain, within particular levels of the chain (referred to as trophic levels), or on the basis of functional groups (e.g., species responsible for mineralization or nitrification processes) rather than taxonomic identity (Peet 1974). Few statements about diversity hold universally. Once an ecosystem has been characterized, however, general claims about the system can be verified or discarded, and the particular significance of changes in diversity can be evaluated. The following points should be emphasized.

(1) A chemical impact is more likely to reduce than increase the number of species within a given trophic level. If several similar species coexist because of specialization at different points of a habitat or resource, heavy mortality affecting various species unevenly can override more subtle ecological differences, thus eliminating some species. The only way in which chemicals can increase the number of species present is by reducing the biota to such an extent that new species can immigrate into the system; but these new species are more likely to displace existing populations that have been drastically reduced in abundance.

(2) Changes in the relative abundance of several species are not predictable. If the most abundant species is also most sensitive to the presence of a chemical, the population size of rarer species could increase and thus equalize the abundance of all. The response of those species not subject to predation is likely to be a change in size of the population. Population changes for species subject to predation, however, can be partially buffered, because decreases in the prey also reduce the number of predators within certain lag-times.

(3) Increased diversity at a given trophic level can lead to (a) smaller variations in total biomass as conditions fluctuate within a range that may or may not be optimal for different species, and (b) greater fluctuations in the composition of species.

(4) The graphic representation of biomass content for some food

chains has a pyramid shape, i.e., the amount of biomass is large at lower levels and small at upper levels. Although a disturbance is usually most severe at these upper levels, extremely toxic substances can reduce species abundance for all. An uneven distribution of effects is possible, however; for example, if a chemical reduces the ability of prey to avoid predation, the result may be an increase in the predator population despite a reduction in prey.

(5) Substantial diversity in communities of microorganisms is more likely to increase degradation of a chemical. If a chemical reduces this diversity, the capacity of an ecosystem to detoxify other pollutants could be reduced.

On the basis of these points, diversity is a system property that is likely to be a sensitive measure of ecosystem contamination.

Productivity and Biomass

Energy and nutrient flows in an ecosystem can be described in terms of either aggregated productivity (i.e., rates of flow-through or turnover) or standing stocks of biomass. Productivity levels can be used to determine rates of possible harvesting for one system that may be transferred to others. Among the factors that influence the amount of biomass within an ecosystem are the capacity of system components to retain water, the presence of structures to prevent loss through air or water runoff, and the capability to dilute the substance. Not all productivity is equally useful, and not all levels of biomass are equally desirable in a system. Whether a change in productivity or biomass proves to be an adverse response depends on properties of the particular system (Terrierer et al. 1966).

In general, a system characterized by high productivity experiences more rapid rates of nutrient or mineral turnover. Sporadic inputs of chemicals into such a system would be processed relatively quickly and might be observed as pulses. A system characterized by low productivity retains the effects of environmental changes much longer; intermittent insults in a slow-moving system would more likely be observed as an average over time.

A change in productivity or biomasss can result from an increase in either mortality or natality of one or more species or from a reduced rate of nutrient flow. Increased mortality shunts more nutrients back to detritus-producing species rather than allowing nutrients to be transferred through the system. The biomass of the affected species may or may not change with increased mortality or natality. Increased decomposition could increase loss of minerals by leaching from terrestrial systems.

Processes of decomposition also affect levels of productivity and biomass. Decomposition of organic materials is an important property of natural ecosystems in that it leads to the conversion of nutrient elements bound in the biomass, detritus, and humus back into the inorganic forms needed for growth of aquatic and terrestrial plants, referred to as mineralization.

In a given ecosystem, mineralization rates can vary greatly throughout the course of a year, depending upon such environmental factors as temperature, oxygen availability, pH, and—in terrestrial systems— moisture level. The vulnerability of primary production to alterations of mineralization activity depends on the susceptibility of microbial populations, the fraction of nutrients supplied to the nutrient pool by mineralization, and the size of the inorganic nutrient pool relative to seasonal nutrient demand of primary producers. Terrestrial plants, for example, derive inorganic carbon from the enormous atmospheric reservoir of CO_2, and return carbon to the system in 10 to 20 years; in many lakes the inorganic phosphorus pool has a return time on the order of days or less (Hayes et al. 1952). Nitrogen as well as phosphorus may cycle rapidly (days or less). Generally a system is more vulnerable to alterations of mineralization rates by toxic substances if turnover times of inorganic nutrient pools are short and if nutrient cycles are closed (e.g., reduced nitrogen fixation or inaccessibility of external nutrient sources).

The assimilation of inorganic elements by microorganisms provides an important sink for nutrients. Because primary producers and animal consumers compete for these nutrients in certain situations, the relative and absolute rates of mineralization and microbial nutrient assimilation can have an important influence on ecosystem behavior. Little information is available, however, about the effects of chemicals on such factors as the mineralization of many nutrient compounds, the microbial assimilation of their inorganic forms, or the ultimate partitioning of increased detritus among plants, detrivores, detritus pool, and inorganic nutrients.

One should not assume that systems characterized by high productivity always exhibit higher rates of mineralization and nutrient turnover than more oligotrophic systems. For example, there is evidence that bacteria in unproductive lakes can significantly degrade many pesticides, utilizing them as organic carbon sources or as sources of nitrogen or phosphorus. These elements are often present at growth-limiting concentrations in oligotrophic systems (C. Goldman, University of California, personal communication, 1980).

Connectivity

Connectivity refers to the intersection of pathways for transferring materials, including chemicals, within the system. High connectivity can disperse a compound throughout a system, and reduce localized effects. This might result in potentially rapid bioaccumulation and subsequent production of toxic effects in secondor third-level consumer organisms. For those compounds that do not accumulate in food chains, high connectivity reduces toxic effects by increasing dilution of the substance. Increased detoxification also can occur if the detoxifying organisms are exposed to a chemical at increasing rates. Each time a compound is transferred to another compartment of the system, physical and chemical properties can be altered. Highly connected systems have the potential to decrease toxic effects for some substances.

A qualitative assessment of connectivity within a system facilitates making predictions about the transfer of potentially toxic effects. This analysis of connectivity can be used in the following ways.

(1) In establishing general principles about the point of impact. For example, effects of a chemical may be concentrated in the upper levels of the food chain (Ribeyre et al. 1979); if a food resource common to several species is reduced, the impact of a substance might be absorbed by the inedible specialized consumers. Although principles based on strictly qualitative evaluations often are not universal, they can indicate potential targets for toxic effects.

(2) As a technique for validating proposed models of interrelationships among components of an ecosystem. For example, a system comprising four components—a nutrient resource (N), two consumers (H_1, H_2), and a predator (P)—may have several component interrelationships beyond those shown in Figure 3.5; H_1 and H_2 can inhibit the release of nutrients that, in turn, can stimulate H_1 and H_2 differentially; the growth of H_2 can be inhibited by P. If the direction and mode of the relationships were not known, the impact of a substance would not be predictable unless representative alternative models were used (see Appendix B for a discussion of such alternative models).

(3) In identifying the point of entry into the ecosystem; correlations between variables can be examined for this purpose. For example, if the consumer (H_1) is the point of entry to the system, a nutrient source (N) and predator (P) can respond in the same manner (increase or

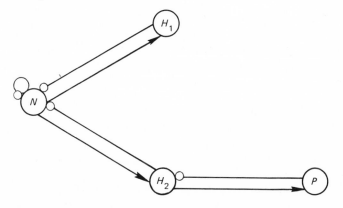

FIGURE 3.5 Four components of a hypothetical ecosystem. (————→ indicates a positive effect and ————O a negative effect.)

decrease) to the disturbance (see Model 14 in Figure B.1, Appendix B). If the disturbance enters via other points, N, P, and H_1 will respond in opposite directions.

Patterns of correlation among different variables for several models are indicated in Table B.1, Appendix B. The important point is that even though a prediction of an increase, no response, or a decrease is weak, the joint confirmation of these correlation patterns can identify the point of entry. Evaluating patterns of correlation can confirm the presumed structure of a system and can be the basis for predicting potential impacts of introduced substances. This approach also allows for the determination of possible responses of a system upon introduction of new substances.

Resistance and Resilience

Ecosystems with high degrees of resilience and resistance can be expected to be less susceptible to disturbance (Westman 1978, Webster and Patten 1979). Resistance suggests the ability of a system to absorb an impact without significant change from normal fluctuations. Highly resistant systems could signify rapid rates of transformation or removal of a substance from the system. It indicates that few organisms are susceptible to a particular impact, or that these organisms are able to rapidly metabolize or detoxify the chemical. Once some threshold is exceeded in a resistant ecosystem, however, the induced change may be quite severe. Resistance to the presence of a chemical can be changed on a seasonal basis, thus altering composition of species and the response of a population or the system to increased mortality.

Chemicals affect ecosystems by changing biological characteristics (e.g., mortality, fecundity, development rate, feeding efficiency, competition coefficients) or by causing a recurrent disturbance of the state of the system. Substances also may alter an ecosystem in any of the following ways.

- *Changes in the abundance of the component species can occur.* The importance of these changes depends on the sensitivities of the species. Accurate predictions of change, in turn, depend on the ability to measure the degree of species sensitivity to the impact of chemicals (Patrick 1949, Patrick et al. 1954).
- *Changes in some function of the system are possible.* A change in mineralization can affect the population size of a key species or change species diversity in the system (Bormann and Likens 1979).
- *Extinction of a species can result.* Destabilization of the system can result in extreme fluctuations in species abundance with extinction resulting.
- *Change in stability of the system may occur.* Change in the behavior of a few species or fluctuations in a functional process will affect stability. Reduced stability increases the time required for a system to return to normal conditions after a disturbance.

Geographic Specificity

The response of an ecosystem to the presence of a chemical depends strongly on particular characteristics of the system. For example, the difference in temperature ranges between ecosystems located at different sites can be a major factor in differences in rates of decomposition between the systems. Geographic location of a terrestrial system also can determine rates of productivity, which are influenced by conditions of underlying rock strata and soil fertility. Responses of a system to a disturbance can be quite local; for example, past instances of fertilizer applications on nearby fields can influence how a particular system might respond to the introduction of a new chemical; a similar system at another location might respond differently. In locations where organisms exist under conditions that are marginal for their survival, the impact of a chemical may be more severe than at locations where optimal conditions exist. The altitude at which a system is found might change the impact of some substances if these are easily transformed by ultraviolet light; snow and ice cover are obvious physical factors that alter the effects of chemicals on ecosystems by greatly reducing rates of degradation, photolysis, and transport. Ecosystems located in areas continuously affected by human activities—nontoxic-waste disposal, soil impaction and disturbance, creation of noise, and harvest of renewable resources such as timber, fish, and wildlife—also would exhibit very site-specific responses to chemicals. Heated effluents also might change the site-specific response to a compound. (For abundant examples see annual reviews of literature in the Journal of the Water Pollution Control Federation.)

Interactions of Populations

Organisms respond to their own population density as well as to that of other species. Social structure and the stimuli required for reproduction and survival are density-dependent factors that might influence a particular population and hence an ecosystem. Toxic effects might directly reduce the size of a population of a particular species to some threshold level (for reproductive success) and indirectly cause extinction of those and other organisms. Although the importance of social structure and stability of interactions to the maintenance of animal populations is not well understood, a great reduction in density could be disruptive in some highly gregarious and socially organized species.

Changes in the density of a species as a result of exposure to toxic chemicals can increase susceptibility of the whole ecosystem to other stresses. If the same or similiar species were not available for recolonization, long-lasting effects in the ecosystem can result. Test methods need to be developed that will permit identification of those changes resulting from a chemical and those that are results of natural fluctuations.

Changes in the interaction of populations after introduction of chemicals can have important effects on the structure of an ecosystem. This relationship is perhaps most obvious in systems having predator and prey, plant and herbivore, or host and parasites. If an ecosystem is to continue to function after the introduction of a compound, it is important that interactions of populations remain intact. Extinction of a preferred prey species, or asynchronous development of a host could lead to a drastic decline in the abundance of the predator or the parasite if alternative prey or hosts are less desirable or not available.

This phenomenon is related not only to those particular interactions but also to the diversity and density of species that compete for common resources. The increase in a number of pest species as a result of introduced chemicals has been well documented (e.g., Van den Bosch 1969).

In addition, effects of a substance (e.g., pesticide) on nontarget organisms often exceed those on target species, especially in the long term; DDT and PCB contamination provide excellent examples.

Artificial manipulation of species in some systems (i.e., stocking an area with a particular animal), particularly intertidal areas, has resulted in major ecosystem effects. For example, Paine (1976) has demonstrated extensive alteration in community structure by manipulating the population size of a predator, *Pisaster*, and its prey, *Mytilus*. Some substances can duplicate effects of artificial manipulations by altering density factors for selected species. If the effect of a substance is selective for one sex, alterations of ratios between males and females may not show measurable effects for several generations. Changes in habitat structure also occur as a result of slight reductions in competitive interactions among closely allied species.

Genetic and Taxonomic Variability

There is ample evidence that the introduction of chemicals into an ecosystem causes selective mortality in a number of species, thereby reducing taxonomic variability of the system (Giles 1970; Stickel 1968, 1975). If this selection process also reduces genetic variation in the population, the response of populations and the system to rapidly changing conditions is impaired. For example, results of studies of blood-protein chemistry in wildlife populations suggest that heterozygous individuals have a greater potential to survive under less than optimal environmental conditions and also may be more successful in reestablishing the population in new habitats (Smith et al. 1976, 1977). Chemicals that reduce heterozygosity can lower the resilience or resistance of an ecosystem to other stress factors.

Although it is rare for a substance to affect a wide spectrum of species in the same manner, chemicals present in the environment can selectively affect the more susceptible species, reducing diversity of the system. The chemical and physical properties of a compound can influence the rate of biological responses in a system. Chemicals that disproportionately influence producer organisms or species responsible for degradation and mineralization will have a serious impact.

Nutrient Retention and Regeneration

Changes in physical and biological diversity of ecosystems by natural or chemical influences are major factors in alterations of nutrient flow and system regeneration. For example, removal of forests from watershed areas and their increased use for agriculture have greatly increased the loss of nutrients largely due to excess water runoff and reduced uptake of these compounds by natural plant communities. When large-scale landscape changes are accompanied by increased use of chemicals, a common occurrence (Winteringham 1979), the impact may be on nutrient cycling in the ecosystem. Chemicals can alter the retention and regeneration of nutrients by several means, including the selective removal of detrivores and primary producers; these alterations affect organisms at higher levels of the food chain, and reduce total productivity, biomass, species diversity, and the ability of the system to recover from the chemical insult.

The ability of a system to recover is related to the properties of the chemical as well as the properties and size of the affected area. If only small patches in the system have been exposed, recovery could be rapid. If larger areas or the entire ecosystem are affected by the substance, recovery could take much longer. This would be particularly true if the chemical eliminated all populations at key points in the structural network of the system. Reestablishment would be slow if there were little immigration of new species or if the substance were persistent. If nutrient losses were excessive and if the original composition of species were dependent on nutrient availability, an affected ecosystem might never recover its former state of complexity. Potential changes in nutrient retention or regeneration of a system resulting from the presence of a chemical should be determined when assessing its impact on an ecosystem.

Composition of Functional Groups

The identification of invertebrate functional groups, as defined by Cummins and Klug (1979), or guilds (see Root 1967) facilitates analyses of resource utilization in aquatic or terrestrial ecosystems. It is a common practice to group invertebrates (Merritt and Cummins 1968) and microorganisms on the basis of functional contributions to the ecosystem (i.e., degradation of a particular substrate such as cellulose or lignin). For example, shifts in a resource base, both quantitative and qualitative, can be identified by changes in the relative biomass abundance and composition for particular feeding groups (e.g., stream macrovertebrate benthic communities; Cummins and Klug 1979). Because functional composition is sensitive to alterations in resource inputs, it is both responsive to toxic effects and predictive of changes in general ecosystem properties, such as nutrient turnover rates. As an example of the latter, the leaf-litter feeders (shredders) are responsible for 20 to 30 percent of the conversion of coarse particulate organic matter to other forms, e.g., fine particles, CO_2, and animal biomass (Petersen and Cummins 1974).

FACTORS INFLUENCING POSSIBLE IMPACTS

Several properties of an ecosystem determine the extent or magnitude of chemical effects. Evaluations of chemical toxicity on ecosystems should include some consideration of the factors discussed in the following sections.

CAPACITY FOR STORAGE AND DETOXIFICATION

There are several pathways for the detoxification and reduction of chemicals. Such mechanisms as decomposition by photolysis and hydrolysis, metabolism by living organisms, transport from one component to another (e.g., soil to water, plants to animals), are reviewed in Chapter 2. Long-term storage of substances in either lethal or sublethal concentrations can result in sudden and intermittent reentry of chemicals into a system, or can effectively remove their environmental threat. Storage of chemicals transported by air in polar icecaps or abyssal depths of oceans would insure their long-term immobility. Although some recirculation is possible over time, little is known about the capacity of such storage sites for recirculation of substances. Other storage sites include sediments in lakes and rivers; however, these may be susceptible to frequent disturbance and thus periodic reintroduction of chemicals into the system.

Long-term deposition provides time for nonbiological degradation of a material or provides an opportunity for transformation by microbial action, either aerobic or anaerobic. Ecosystems that have a large capacity for long-term storage are capable of absorbing significant amounts of chemicals and reducing their impact.

ADAPTIVE POTENTIAL

That populations can adapt rapidly to changing natural conditions is suggested by herbivore-host plant relationships (Whittaker and Feeney 1971), and there is also evidence that genetic selection in natural systems results from the impact of chemicals (Ferguson et al. 1966, Ferguson and Bingham 1966, Wagner 1974). Most notable examples are insect and some vertebrate resistance to pesticides and the occurrence of "industrial melanism" in moths in Great Britain (Kettlewell 1955). When assessing the potential environmental consequences of the presence of chemicals, adaptation may be an important factor to consider for the following reasons.

(1) Short-term tests may overestimate the impact in the long run.

(2) Because species have unequal capacities to respond to disturbances in their environment, those that can adapt quickly may eventually dominate the system, thus reducing diversity.

(3) The mode of adaptation may involve reduced rates of growth, lower fecundity, tolerance to stress, or altered food and habitat preferences that can change competitive relationships within the ecosystem.

(4) Genetic changes in the frequencies of biochemical markers may be indicators of differential mortality or fecundity but these changes might be difficult to evaluate.

Observed responses to selection pressures can provide a lower bound on estimates of selection intensity. If individual ages are determined, the intensity of selection at each age class could be estimated. Because only differential rates are noted, such studies might underestimate mortality, and because this method registers all causes of death, analyses of age structure could overestimate potential toxicity of chemicals.

Many factors can influence the magnitude of chemical effects. Length of life cycle may alter the modes of genetic selection to chemicals. Selection for enhancement of metabolism and rapid excretion could result in lowered toxic effects within populations as well as fewer genetic changes in long-lived species, especially if reproductive processes are affected.

The actual mechanism for selection as a result of exposure to a pollutant, however, is largely unknown. Physiological processes, behavioral changes, chemical/physical effects on the nervous system, or direct changes in the genetic processes themselves are involved. Rates of mutation in some unicellular organisms are known to increase as a result of exposure to pollutants (Ames 1979). Any chemical introduced into a natural system can alter the normal basis for selection, possibly changing genetic structure, and may result in a system that is more or less resistant to future introduction of similar chemicals.

TEMPORAL CHANGE

Temporal change is an important part of every functioning system, even those that are considered highly stable. In fact, perceptions of ecosystem stability are largely time dependent. For example, temporal changes in the density of a species resulting from seasonally induced recruitment and mortality require that the system undergo change to accommodate any increase or decrease in population size. Changes in ecosystem function frequently are related to seasonal shifts in climatic patterns, temperature, and perhaps most importantly, rainfall. The introduction of a chemical during a period of climatic extremes could result in a much greater effect than if exposure occurred at a more "normal" time of the year. Because time-related changes in physical processes of the system (e.g., energy flow, nutrient and mineral loss or gain) are integrated responses, measurement of changes in them may indicate effects due to the presence of a chemical. Seasonal shifts in these processes may not be great and baseline measurements could be more easily determined.

Rates of immigration and emigration also influence cyclical processes in ecosystems. The sudden influx of several million roosting birds, for example, can change the rate of energy flow and nutrient accumulation within a system which, in turn, may influence the response of the system to environmental contaminants. The impact of chemicals on other sequential events, such as delays in reproductive cycles, also should be evaluated. Temporal changes of varying duration often influence basic population or ecosystem characteristics and interrelationships (e.g., recruitment, mortality, and nutrient cycling), therefore effects of chemicals must be evaluated in light of potential changes over time.

CHANGES IN SPECIES INTERACTIONS

The alteration of species interactions resulting from changes in habitat conditions (i.e., temperature or moisture) can influence the effect chemicals will have on the ecosystem. Major shifts in species diversity could enhance or reduce the effect of contaminants. A study of predator-prey relationships in Lake Washington illustrates this point. Mysid crustaceans (*Neomysis*) apparently are key predators in the lake. The species selectively prey on *Daphnia* in preference to copepods and at certain times can eliminate or almost eliminate *Daphnia* from the zooplankton community. This has repercussions on pollutant levels of the ecosystem because *Daphnia* is a more effective feeder than the other zooplankton. During the summer after substantial decreases in populations of *Neomysis*, *Daphnia* become the dominant genus and the result is increased transparency of the lake water (Edmondson 1979).

Exploitative competition is another important factor that can influence the magnitude of environmental contamination. Selected examples of competitive exclusion in simplified laboratory systems are presented below. They illustrate that the outcome of competition can be affected either by (1) the presence of other organisms or (2) the physical conditions of the habitat. If a chemical affects species differentially, then competitive interactions and the outcome under various conditions must be considered before accurate predictions of effects can be made.

1. Studies of the grain beetle *Tribolium* have provided useful information on species interaction, although the competitive mechanism is exceedingly complex, involving both antibiosis and predation. In competition experiments with *Tribolium confusum* and *T. castaneum*, the former succeeds if test conditions include low temperature and moisture levels, but does not succeed under hot, moist conditions. At intermediate temperatures and moisture, the outcome is uncertain, at least with genetically heterogenous populations. If closely inbred stocks are used, the results for intermediate conditions depend on the stock tested, certain strains of *T. confusum* always outcompeting certain stocks of *T. castaneum* and vice versa (Lerner and Ho 1961). The pattern of fluctuations in population size and the outcome of competition for *Tribolium* also is affected by the presence of the parasite *Adelina*, as one species of *Tribolium* is much more resistant to parasitism than others.

2. Another example involves *Paramecium*. The success of *Paramecium caudatum* over *P. aurelia* depends on the type of food source provided (Gause 1934).

3. The competition for food between bluegill and trout is strongly affected by temperature. Bowen and Coutant (1971) measured the proportion of food eaten by each and found that small differences in temperature have significant effects. At a given initial temperature the trout consumed 75 percent of the available food; after an increase of 3°C the amount eaten by the trout decreased to 25 percent.

In each of these examples, the effect of a chemical might be different depending on which species is most successful at the time the system is contaminated. Several hypothetical outcomes may need to be tested before potential toxicity of a chemical can be fully evaluated.

SPATIAL DISTRIBUTION

The satisfactory development of an individual's niche is dependent on spatial relationships with other organisms. Spatial relationships are important not only for a given species, but also for closely associated codominant species. Chemical impacts that alter population density can change spatial relationships and, at certain times of the year, such changes may be critical to the reproductive process (e.g., certain birds require stimulation by a neighbor before successful development of the reproductive sequence occurs, nest-building, copulation, and incubation).

Spatial relationships also play a critical role in the physical transfer of contaminants in ecosystems and thus influence the impact of the contaminants. Changes in the density and spatial distribution of invertebrates that are important to soil texture and structure could alter the basic characteristics of soil. Physical transport of chemicals is enhanced by rapid movement of water to lower soil horizons, thus resulting in transfer of the contaminant to storage sites and rapid decontamination of the soil surface.

Biological transfer can be influenced by spatial distribution of organisms, including migratory birds, insects, predators and their prey, and plants and their pollinators. In some cases, however, the transfer of chemicals might be inhibited by sedentary species that accumulate large amounts of the substance. Spatial distribution of organisms in aquatic systems often determines the rate of chemical uptake and ultimate deposition of substances in deep ocean sediment.

Spatial distribution obviously is related to population density, which is strongly influenced by immigration patterns and mortality of individuals. If these change as a result of exposure to chemicals, the spatial relationship also changes. Threshold effects at both low and high densities could be caused by alteration of the status of a population within the system. If the impact of a compound is highly species-specific and reduces the population size of one competitive species, large increases in populations of another might occur, resulting in high-density effects such as social stress or over-utilization of resources. If a population is greatly reduced, low densities could result in extinction. Although changes in spatial distributions of individuals and populations are important responses to the introduction of a chemical into an ecosystem, much research is yet needed to develop suitable techniques for assessing changes related to chemical contamination.

DENSITY DEPENDENCE

Although ecologists do not agree on which ecological variables are density dependent, physical partitioning of a substance within an ecosystem can be related to the abundance of organisms present. If a toxic substance is introduced as a single release in an open system with a very dense population, the potential exists for greater dilution and reduced uptake per individual (Terrierer et al. 1966). Static tank tests with fish have corroborated this effect. In a complex and highly dense system, the biological and physical pathways of distribution or accumulation also vary, thus reducing the impact of a chemical on a system.

Populations of very high or low densities usually are less stable and more easily disturbed by added environmental or chemical stress. The effects of such stress can be either additive or synergistic. If extreme natural factors are tolerated by the component species of the system, the potential for a large impact is increased. This also could occur at lower densities if mostly physical rather than biological factors are involved. Some species require minimum threshold densities before complex social interactions can lead to successful reproduction. Thus, if a pollutant lowers density below this level, serious declines or actual extinction of species can result. This occurs even without observed increases in mortality if the complex behavioral patterns that lead to changes in birth rates are disturbed.

The toxicological properties of any chemical influence its density-dependent impact. A highly toxic substance that induces heavy mortality removes susceptible individuals regardless of density. A less toxic substance can be partitioned throughout the system such that mortality is not the major impact; density of the system can be involved directly in this partitioning process. Not only might the impacts change as a result of partitioning, but in those instances where metabolic activity is important for detoxification, high density results in a more rapid detoxification of the substance. Changes in population density are not always linear (as illustrated in the following examples):

(1)   death rates that increase exponentially as population size increases, because of mutual poisoning or other crowding effects (Verhulst effect);
(2)   reproductive rates that decrease rapidly during times of low density, because of the difficulty of finding mates (Allee effect); and
(3)   predation rates that fall to zero, or at least to sharply reduced values, when densities of prey are sufficiently low.

Theoretically, there is little doubt that density dependence exerts an important influence on the stability of ecosystems. For example, increasing the relative strength of a nonlinear loss rate in a population is likely to render most mathematical models more unresponsive to externally imposed disturbances. Steele (1974) has emphasized the stabilizing role of a density-dependent rate of predation. An Allee effect results in instability, thus increasing the opportunity for extinction of a sparse population.

The measurement of density dependence also is of interest in studies of succession of microorganisms. Luckinbill (1978) has demonstrated that pure cultures of *Escherichia coli* grown under density-dependent controls are superior competitors to cultures adapted to conditions permitting log-phase growth, whether the comparison is made under crowded or uncrowded conditions. Density-dependent factors influence not only the physical partitioning of a substance but also relate to the responsiveness of the individual and the population to it.

Whether density dependence is an empirically accessible concept is a controversial issue (Ehrlich and Birch 1967, Lidicker 1978, Slobodkin et al. 1967). Much of the debate has occurred through attempts to identify and quantify density dependence using correlation analysis; these analyses examine the change in a population over a fixed period of time to determine if that change has a nonlinear relationship to population size. This approach, however, is beset with statistical traps (Eberhardt 1970). Because of the potential importance of density dependence as an indicator of change or stability, the development of reliable measurement techniques is needed.

SUMMARY

Certain characteristics of populations and ecosystems are vulnerable to the presence of a chemical and should be considered in any evaluation of potential impacts. Pertinent characteristics of populations include changes in mortality, fecundity, growth rates, age distribution, and phenotypic variation. Interactions of stress factors, behavioral patterns, and migration are attributes of individual organisms, but at some future time they may be described for entire populations and ecosystems as well.

Methods should be developed for measuring changes in such system properties as diversity, levels of productivity and accumulation of biomass, degree of connectivity, resistance and resilience, interaction of species, taxonomic variability, flow of energy and essential nutrients, and composition of functional groups. Ecosystem responses to a chemical will be influenced by such factors as the capacity of the system to store or detoxify the chemical, the adaptive potential of species, natural fluctuations in abundance of species and life-cycle periodicity, temporal changes in climate, alterations in the interactions of species due to natural conditions, spatial distribution of components in the system, and density dependence.

All these attributes (whether of ecosystems, populations, or individuals) are affected by the presence of chemicals, and further effort is needed to develop suitable methods for quantifying those changes that are consequences of the presence of chemicals. A critical facet of assessment procedures is the separation of changes in these attributes as a result of natural fluctuations from changes caused by the chemical of concern.

REFERENCES

Ames, B.N. (1979) Identifying environmental chemicals causing mutations and cancer. Science 204(4393):587-593.

Azevedo, J.A., Jr., E.G. Hunt, and L.A. Woods, Jr. (1972) Melanistic mutant in ring neck pheasants. Calif. Fish Game 58(3):175-178.

Bardach, J.E., J.H. Todd, and R. Crickmer (1967) Orientation by taste in fish of the genus *Ictalurus*. Science 155:1276-1278.

Batterton, J.C., G.M. Boush, and F. Matsumura (1972) DDT inhibition of NaCl tolerance by bluegreen alga, *Anacystis nidulans*. Science 176(4039):1141-1143.

Biesinger, K.E. and G.M. Christensen (1972) Effects of various metals ton survival, growth, reproduction, and metabolism of *Daphnia magna*. J. Fish. Res. Board Can. 29:1691-1700.

Bormann, F.H. and G.E. Likens (1979) Pattern and Process in a Forest Ecosystem: Disturbance, Development, and the Steady State Based on the Hubbard Brook Ecosystem Study. New York, N.Y.: Springer-Verlag New York, Inc.

Bowen, S.H. and C.C. Coutant (1971) Thermal effect on feeding competition between rainbow trout and bluegill. Third National Symposium on Radioecology, Vol. 1. U.S. Atomic Energy Commission, Oak Ridge National Laboratory, May 10-12, 1971. ORNL Publication Number 550. Oak Ridge, Tenn.: Oak Ridge National Laboratory.

Brungs, W.A. (1969) Chronic toxicity of zinc to the fathead minnows, *Pimephales promelas* Rafinesque. Trans. Am. Fish. Soc. 98(2):272-279.

Bullock, T.H. (1973) Seeing the world through a new sense: Electromagnetism in fish. Am. Sci. 61(3):316-325.

Canton, J.H., P.A. Greve, W. Slooff, and G.J. Van Esch (1975) Toxicity accumulation and elimination studies of alpha-hexachlorocyclohexane (alpha-HCH) with freshwater organisms of different trophic levels. Water Res. 9:1163-1169.

Cole, H.A., ed. (1979) The Assessment of Sublethal Effects of Pollutants in the Sea. London: Royal Society of London.

Comita, G.W. (1956) A study of a calanoid copepod population in an Arctic lake. Ecology 37:576-591.

Comita, G.W. (1972) The seasonal zooplankton cycles: Production and transformation of energy in Severson Lake, Minnesota. Arch. Hydrobiol. 70:14-66.

Cooke, A.S. (1971) Selective predation by newts on frog tadpoles treated with DDT. Nature 229:275-276.

Cummins, K.W. and M.J. Klug (1979) Feeding ecology of stream invertebrates. Annu. Rev. Ecol. Syst. 10:147-172.

Cunningham, P.A. and D.S. Grosh (1978) A comparative study of the effects of mercuric chloride and methyl mercury chloride on reproductive performance in the brine shrimp. *Artemia salina*. Environ. Pollut. 15:83-99.

Eberhardt, L. (1970) Correlation, regression, and density dependence. Ecology 51(2):306-310.

Edmondson, W.T. (1979) Lake Washington and the predictability of limnological events. Arch. Hydrobiol. 13:234-241.

Ehrlich, P.R. and L.C. Birch (1967) The 'balance of nature' and 'population control.' Amer. Natur. 101:97-107.

Farr, J.A. (1978) The effect of methyl parathion on predator choice of two estuarine prey species. Trans. Am. Fish. Soc. 107(1):87-91.

Fendley, T.T. and I.L. Brisbin (1977) Growth curve analyses: A potential measure of the effects of environmental stress upon wildlife populations. Pages 337-350, Transactions of the 13th International Congress on Game Biology, edited by T.J. Peterle. Washington, D.C.: The Wildlife Society and Wildlife Management Institute.

Ferguson, D.E. and C.R. Bingham (1966) The effects of combinations of insecticides on susceptible and resistant mosquito fish. Bull. Environ. Contam. Toxicol. 1:97-103.

Ferguson, D.E., J.L. Ludke, and G.G. Murphy (1966) Dynamics of endrin uptake and release by resistant and susceptible strains of mosquito fish. Trans. Am. Fish. Soc. 95:335-344.

Friend, M. and D.O. Trainer (1970) Polychlorinated biphenyl: Interaction with duck hepatitis virus. Science 170:1314-1316.

Friend, M., M.A. Haegele, and R. Wilson (1973) DDE: Interference with extra-renal salt excretion in the mallard. Bull. Environ. Contam. Toxicol. 9(1):49-53.

Gause, G.F. (1934) The Struggle for Existence. Baltimore: Williams and Wilkins.

Giles, R.H., Jr. (1970) The ecology of a small forested watershed treated with the insecticide malathion—S^{35}. Wildlife Monograph No. 24. Washington, D.C.: The Wildlife Society.

Goodyear, C.P. (1972) A simple technique for detecting effects of toxicants or other stresses on a predator-prey interaction. Trans. Am. Fish. Soc. 101:367-370.

Gould, J.L. (1980) The case for magnetic sensitivity in birds and bees (such as it is). Am. Sci. 68(3):256-267.

Haegele, M.A. and R.H. Hudson (1973) DDE effects on reproduction of ring doves. Environ. Pollut. 4:53-57.

Halbach, U. (1974) Quantitative Beziehungen zwischen Phytoplankton und der Populations Dynamik des Rotators *Brachionus calyciflorus* Pallas. Befunde aus Laboratoriumsexperimenten und Freilandunter-suchungen. Arch. Hydrobiol. 73:273-309.

Hansen, D.J. (1969) Avoidance of pesticides by untrained sheepshead minnows. Trans. Am. Fish. Soc. 98:426-429.

Hasler, A.D. and W.J. Wisby (1949) Use of fish for the olfactory assay of pollutants (phenols) in water. Trans. Am. Fish. Soc. 101:346-350.

Hayes, F.R., J.A. McCarter, M.L. Cameron, and D.A. Livingstone (1952) On the kinetics of phosphorous exchange in lakes. J. Ecol. 40(1):202-216.

Heath, R.G., J.W. Spann, and J.F. Kreitzer (1969) Marked DDE impairment of mallard reproduction in controlled studies. Nature 224(5214):47-48.

Henny, C.J. (1972) An Analysis of the Population Dynamics of Selected Avian Species with Special Reference to Changes During the Modern Pesticide Era. Wildlife Research Report No. 1, U.S. Fish and Wildlife Service. Washington, D.C.: U.S. Department of the Interior.

Henny, C.J. (1977) Birds of prey, DDT, and tussock moths in the Pacific Northwest. Pages 397-411, Transactions of the 42nd North American Wildlife and Natural Resources Conference, edited by K. Sabol. Washington, D.C.: Wildlife Management Institute.

Hutchinson, G.E. (1978) An Introduction to Population Ecology. New Haven, Conn.: Yale University Press.

Jones, B.F., C.E. Warren, C.E. Band, and P. Dondoroff (1956) Avoidance reactions of salmonid fishes to pulp mill effluents. Sewage Ind. Wastes 28(11):1403-1413.

Kania, H.J. and J.O. O'Hara (1974) Behavioral alterations in a simple predator-prey system due to sublethal exposure to mercury. Trans. Am. Fish. Soc. 103(1):134-136.

Keeton, W.T. (1974) The mystery of pigeon homing. Sci. Am. 231(6):96-107.

Kettlewell, H.B.D. (1955) Selection experiments in industrial melanism in the Lepidoptera. Heredity 9:323-342.

Lerner, I.M. and F.K. Ho (1961) Genotype and competitive ability of Tribolium species. Am. Nat. 95:329-343.

Lewis, F.G. III and R.J. Livingston (1977) Avoidance of bleached Kraft pulpmill effluent by pinfish (*Lagodon rhomboides*) and gulf killifish (*Fundulus grandis*). J. Fish. Res. Board Can. 34:568-970.

Lidicker, W.Z. (1978) Regulation of numbers in small mammal populations: Historical reflections and a synthesis. Pages 122-141, Populations of Small Mammals Under Natural Conditions, D.P. Snyder, ed., Vol. 5, The Pymatuning Symposia in Ecology, University of Pittsburgh Press.

Lindauer, M. and H. Martin (1968) Die schwereorientierung den biemen unter dem einfluss des erdmognet feldes. Z. verlag. Physiol. 60:219-243.

Luckinbill, L.S. (1978) R and K selection in experimental populations of *Escherichia coli*. Science 202:1201-1203.

Merritt, R.W. and K.W. Cummins, eds. (1968) Introduction to the Aquatic Insects of North America. Dubuque, Iowa: Kendall/Hunt.

Moore, B.R. (1980) Is the homing pigeon's map geomagnetic? Nature 285(5760):69-70.

National Research Council (1980) Urban Pest Management. Committee on Urban Pest Management, Environmental Studies Board, Commission on Natural Resources. Washington, D.C.: National Academy Press.

Paine, R.T. (1976) Size-limited predation: An observational and experimental approach with the *Mytilus-Pisaster* interaction. Ecology 57(5):858-873.

Patrick, R. (1949) A proposed biological measure of stream conditions, based on a survey of the Conestoga Basin, Lancaster County, Pennsylvania. Pages 277-341, Proceedings: Academy of Natural Sciences of Philadelphia, Vol. 101. Philadelphia, Pa.: Academy of Natural Sciences of Philadelphia.

Patrick, R., M.H. Hohn, and J.H. Wallace (1954) A New Method for Determining the Pattern of the Diatom Flora. Pages 1-12, Notulae Naturae. No. 259. Philadelphia, Pa.: Academy of Natural Sciences of Philadelphia.

Peet, R.K. (1974) The measurement of species diversity. Annu. Rev. Ecol. Syst. 5:285-307.

Peterson, R.O. (1977) Wolf ecology and prey relationships on Isle Royal. National Park Service Scientific Monograph Series No. 11. Washington, D.C.: National Park Service.

Peterson, R.O. and K.W. Cummins (1974) Leaf processing in a woodland stream. Freshwater Biol. 4:343-368.

Presti, D. and J.D. Pettigrew (1980) Ferromagnetic coupling to muscle receptors as a basis for geomagnetic field sensitivity in animals. Nature 285(5760):99-101.

Ribeyre, F.A., A. Boudou, and A. Delarche (1979) Interest of the experimental trophic chains as ecotoxicological models for the study of the ecosystem contamination. Ecotoxicol. Environ. Saf. 3:411-427.

Roberts, J.R., D.W. Rodgers, J.R. Bailey, and M.A. Rorke (1978) Polychlorinated Biphenyls: Biological Criteria for an Assessment of Their Effects on Environmental Quality. Ottawa, Canada: National Research Council Canada.

Root, R.B. (1967) The niche exploitation pattern of the blue-gray gnatcatcher. Ecol. Monog. 37:317-350.

Slobodkin, L.B., F.E. Smith, and N.G. Hairston (1967) Regulation in terrestrial ecosystems and the implied balance of nature. Am. Nat. 101:109-124.

Smith, M.H., H.O. Hillestad, M.N. Manlove, and R.L. Marchinton (1976) Use of population genetics data for the management of fish and wildlife populations. Pages 119-133, Transactions of the 41st North American Wildlife and Natural Resources Conference, edited by K. Sabol. Washington, D.C.: Wildlife Management Institute.

Smith, M.H., H.O. Hillestad, M.N. Manlove, D.O. Straney, and J.M. Dean (1977) Management implications of genetic variability in loggerhead and green sea turtles. Pages 302-312, Transactions of the 13th International Congress on Game Biology, edited by T.J. Peterle. Washington, D.C.: The Wildlife Society and Wildlife Management Institute.

Southern, W.E. (1974) The effects of superimposed magnetic fields on gull orientation. Wilson Bull. 86(3):256-271.

Steele, J.H. (1974) The Structure of Marine Ecosystems. Cambridge, Mass.: Harvard University Press.

Stickel, L.F. (1968) Organochlorine Pesticides in the Environment. Special Scientific Report on Wildlife, No. 119. U.S. Fish and Wildlife Service. Washington, D.C.: U.S. Department of the Interior.

Stickel, W.H. (1975) Some effects of pollutants in terrestrial ecosystems. Pages 25-74, Ecological Toxicology Research, edited by A.D. McIntyre and C.F. Mills. New York: Plenum Publishing Corporation.

Sullivan, J.F., G.J. Atchison, D.J. Kolar, and A.W. McIntosh (1978) Changes in the predator-prey behavior of fathead minnows (*Pimephales promelas*) and largemouth bass (*Micropterus salmoides*) caused by cadmium. J. Fish. Res. Board Can. 35:446-451.

Tagatz, M.E. (1976) Effect of mirex on predator-prey interaction in an experimental estuarine ecosystem. Trans. Am. Fish. Soc. 105:546-549.

Terrierer, L.C., U. Kiigenagi, A.R. Gerlach, and R.L. Borovicka (1966) The persistence of toxaphene in lake water and its uptake by aquatic plants and animals. J. Agric. Food Chem. 14:66-69.

Treshow, M. (1975) Interaction of air pollutants and plant disease. Responses of Plants to Air Pollutants, edited by J.B. Mudd and T.T. Kozlowski. New York: Academic Press.

U.S. Environmental Protection Agency (1975) DDT: A Review of Scientific and Economic Aspects of the Decision to Ban Its Use as a Pesticide. Prepared for the Committee on Appropriations, U.S. House of Representatives, Report No. EPA-540/1-75-022. Springfield, Va.: National Technical Information Service.

Van den Bosch, R. (1969) The toxicity problem-comments by an applied insect ecologist. Pages 97-112, Chemical Fallout: Current research on persistent pesticides, edited by M.W. Miller and G.G. Berg. Springfield, Ill.: C.C. Thomas Publishers.

Wagner, R.H. (1974) Environment and Man. New York: W.W. Norton and Company.

Walcott, C. (1974) The homing of pigeons. Am. Sci. 62(5):542-552.

Walcott, C., J.L. Gould, and J.L. Kirschvink (1979) Pigeons have magnets. Science 205(4410):1027-1029.

Webster, J.R. and B.C. Patten (1979) Effects of watershed perturbation on stream potassium and calcium dynamics. Ecol. Monog. 49:51-72.

Weiss, C.M. (1959) Response of fish to sub-lethal exposures of organic phosphorus insecticides. Sewage Ind. Wastes 31(5):580-593.

Westman, W.E. (1978) Measuring the inertia and resilience of ecosystems. Bioscience 28(11):705-710.

Whittaker, R.H. and P.P. Feeney (1971) Allelochemics: Chemical interactions among species. Science 171:757-770.

Wiltschko, W. and R. Wiltschko (1972) Magnetic compass of European robins. Science 176:62-64.

Winner, R.W., T. Kelling, R. Yaeger, and M.P. Farrell (1977) Effect of food type on the acute and chronic toxicity of copper to *Daphnia magna*. Freshwater Biol. 7(4):343-349.

Winteringham, F.P.W. (1979) Agrochemical-chemical interactions and trends. Ecotoxicology and Environmental Safety 3:219-235.

4 Suitable Test Systems

Emphasis on ecotoxicology has developed only recently, and two separate fields of inquiry have yet to be integrated into standard testing procedures: studies on interactions of species and studies on fates and effects of chemicals. Once data on chemical and physical properties of a chemical are available and incorporated into models of ecosystems, estimates of the environmental fate of the substance can be made (e.g., rapid degradation, accumulation primarily in soil, or transfer through aquatic food chains). Such estimates will be useful in identifying those properties and processes of ecosystems that are most likely to be impaired and in selecting test conditions. After developing test methodology for the properties discussed in Chapter 3, it should be possible to use several test systems to evaluate the impact of a chemical in ecosystems. This chapter will discuss advantages and limitations of various systems.

Before identifying specific test systems, however, criteria must be established with which suitable test procedures can be evaluated. A generally suitable system would satisfy several criteria. Among these are the following.

(1) The test system chosen should have the capability to predict fates and effects for a wide range of chemicals, i.e., the test system cannot be designed for only a specific chemical. Predictions should be possible for effects of variable doses on both individual components and on system-wide functions under a variety of environmental conditions. The

data resulting from a test system should include a range of concentration-response relationships within realistic time limits and for various points of impact.

(2) The test procedures should be verified as resembling the natural system either by using data in mathematical models to predict effects or by validating with field studies; the use of both approaches is preferred.

(3) The results of a satisfactory test system should be capable of replication in other laboratories.

(4) The test procedures should have a sound statistical basis. Source material should be sufficiently abundant to ensure statistically justified sample sizes, and the data generated should be compatible with accepted techniques of statistical analysis.

(5) It should be possible to standardize test procedures, to ensure data compatability among laboratories.

(6) A suitable test system should be as realistic as possible for both conditions of the ecosystem and for the environmental form of the chemical. A realistic system would provide the quality of data needed to make reliable predictions about either a general range of environmental conditions (e.g., fate in southwestern deserts and northern tundra) or specific types of ecosystems (e.g., effects in Lake Ontario or a small inland lake). Realistic behavior of components in laboratory tests should be verified for both a disturbed and undisturbed situation. A test procedure will be realistic if it resembles a particular natural system from which test components were taken, or if test results can be extrapolated to a general class of systems with the use of models.

(7) The test procedure should be as economical and uncomplicated as possible while retaining realistic characteristics. Although ecosystems are complex entities, it should not be inferred that the test systems themselves must also be complex.

In the following sections various types of tests are described and evaluated in terms of the above criteria. Test systems are available that simulate responses of single species, interactions of populations, or properties of ecosystems. It is not the intent of this chapter to discuss test methodologies. Most of the concepts presented in the previous chapter require further research before suitable test procedures can be developed. Recent publications do identify some methodologies that are currently available (Alevras et al. 1979, 1980; Hammons 1980, Larimore and McNurney 1980, Larimore et al. 1980, Logan et al. 1980). No attempt is made in this chapter to specify suitable test species, because choice of them would depend on several factors, including expected fate of the test chemicals and ecosystems of concern. An early study of this topic discussed briefly some species that might be considered in eco-system tests (NRC 1975).

SINGLE-SPECIES TESTS

A discussion of single-species tests is presented in Chapter 1. Although limited in predictive capability, sensitivity, and realism, data from these tests can be valuable as a component of an assessment strategy. Single-species tests, as currently conducted, involve exposing a few individuals to a range of chemical concentrations over variable time periods and under standard laboratory conditions. Because of the small size of the test population, results of these tests should not be considered as representing responses of a population. Results from this class of tests, however, are important in developing test designs of ecological systems; for example, data from range-finding tests can be used to develop concentrations for future tests.

Single-species tests have some distinct advantages: the procedures are simple and can be standardized; the results can be readily verified and replicated (Draggan 1978). The utility of a single-species test, however, is a direct function of the criteria used to select the test organism. These tests are not sensitive to interactions among species or to indirect influences that can dominate in complex ecosystems. Furthermore, the chemical substance has a mode of entry into and behavior in a test situation that are significantly different from natural conditions (see Chapter 1). For these reasons ecosystem effects cannot be predicted using *only* data from single-species tests.

POPULATION TESTS

Many plants, animals, and microorganisms have been ignored for use in laboratory tests in favor of standard sets of species that have been traditionally used in evaluating effects on human health and environmental quality. The standard sets of organisms are not often representative of ecosystems, therefore samples from relevant and ecologically realistic populations should be maintained and tested in laboratories using field conditions, whenever possible.

Parameters that should be monitored for determining the impact of chemicals on populations include some of the best-tested and most reliable measures in studies of population dynamics (May 1976). They include population characteristics used to construct actuarial life tables and survivorship curves (Sharitz and McCormick 1973), and changes in population gene pools, migratory behavior patterns, and food preferences (Davis 1974). The strength of these parameters lies in the fact that the procedures can be replicated, standardized, and verified. The population properties discussed in Chapter 3 are also important indicators of chemical impacts, but additional research will be needed before they can be incorporated into suitable test procedures. Tests using population dynamics lack some realism, and because of the absence of interactions with other species, their predictive capability is limited.

MULTI-SPECIES TESTS

An appropriate testing strategy requires that tests be conducted to determine the effects that interactions of species can have on properties of ecosystems. Conclusive evidence has been obtained to support the hypothesis that chemicals significantly alter interrelationships among species to the extent that the survival of one can be seriously jeopardized (Farr 1977; Zaret 1972, 1975; Zaret and Kerfoot 1975). This type of effect can be detected only with multi-species tests.

Some tests using concepts of predation and competition have been developed (Hammons 1980), but virtually no test development is under way on other types of interactions, including symbiotic relationships, host-plant relationships, parasitism, or social interactions. In addition, most available testing protocols have been developed for aquatic systems, with little attention given to terrestrial ecosystems. Several multi-species systems have been developed to verify theories of ecology and are suitable for use in assessments of impacts due to chemicals. These include procedures for testing competition of aquatic plants (Fisher et al. 1974, Mosser et al. 1972), competition of terrestrial plants (McCormick et al. 1974), predation (Zaret 1972, 1975; Zaret and Kerfoot 1975), and animal behavior (Farr 1977).

The strengths of multi-species tests are their capacity to predict a wide range of phenomena, the potential for standardization of test procedures, and the relatively simple techniques employed. To the extent that population characteristics are subject to indirect influences within ecosystems, realism and breadth of predictive capability are compromised somewhat with these tests. Species interactions, such as predator-prey relationships, however, do provide information on transfer coefficients that is useful in predicting the fates and probable effects of chemicals.

A variety of gnotobiotic systems are suitable for tests of interactions of species. These tests have often been misidentified as microcosms, but they are not intended to be reduced scale models of natural ecosystems—an accepted definition of a microcosm. When we are studying actions of chemicals on an organism, a gnotobiotic system has the disadvantage that the test environment is quite different from natural conditions, and even from conditions in microcosms. The group of species used in these systems is brought together in a way that bears little resemblance to natural succession, and no measure of stress can be made. Because stress from unnatural living environments may act synergistically with the stress related to the presence of a test chemical, descriptions of these factors are essential if the gnotobiotic approach is to be of benefit in ecotoxicology. A number of ecotoxicological problems, however, are not suited for study in gnotobiotic systems. For example, investigation of the effects of chemicals on mineralization requires a realistic detritus pool as well as a wide and representative mix of detritivores. This would be impossible to achieve under gnotobiotic conditions.

ECOSYSTEM TESTS

Tests of the fates and effects of chemical substances on properties of ecosystems comprise three types: laboratory microcosms, field enclosures (e.g., greenhouses), and tests in natural ecosystems. The types are discussed in the sections that follow.

LABORATORY MICROCOSMS

Microcosms are defined as samples from natural ecosystems housed in artificial containers and kept in a laboratory environment. These systems are generally initiated by taking whole samples from ecosystems into the laboratory.

The potential advantages of these test systems are numerous. Perhaps the most important is that effects beyond the level of single species can be identified. Their compactness and common environmental conditions permit both replication and standardization, at least in principle. Uniform conditions facilitate comparisons of different substances. The lack of complicated spatial heterogeneity allows for an accurate definition of physical, chemical, and biological characteristics. Because of the absence of complicating environmental variability, causal relationships are more easily analyzed in microcosms than in natural systems. Different physical, chemical, and biological variables can be tested with minimal effort and with no greater expense than that associated with well-designed single-species tests. Potentially dangerous test substances and radiotracers can be administered without contamination of the natural environment. (Although contaminated microcosm materials must be disposed of eventually, the materials can be placed in carefully selected depositories rather than released into general waste disposal systems.) Rapid evaluation of an impact is often possible.

Several limitations are inherent in microcosms. They are intentional simplifications of natural systems. In particular, the physical environment of a microcosm can be often very different from that of the natural system. As a result, the responses observed must be extrapolated to those expected in nature. Some significant aspects of a natural system (e.g., invaders) are absent. Missing components often can exert crucial influences on responses of organisms to stress factors within the natural habitat. As a microcosm becomes more complex, including more of the biotic and abiotic components of its natural counterpart, this problem is reduced. But residual differences always remain, even between the most complex model system and its natural counterpart.

The small size (up to 1000 l) of most microcosms introduces unavoidable problems of scale that further reduce their ecological realism. These problems are most apparent in aquatic systems (Dudzik et al. 1979; Harte et al. 1980; Jassby et al. 1977a, 1977b; Whittaker 1961). The shallow depths of most aquatic microcosms result in unrealistic influences by benthic compartments on nutrient fluxes and decomposition activities. In marine systems, an attempt has been made to solve this problem. Benthic chambers have been designed to reduce the surface area of sediment that is in contact with the surrounding water (Perez et al. 1977). System realism apparently is enhanced by this design. Shallow depths of microcosms also distort the vertical migration patterns of zooplankton and the loss of phytoplankton as they move from the water column to the sediment layer.

Although inclusion of larger organisms (e.g., fish, snails, and larger crustacea) is desirable, these species can overwhelm nutrient cycles (Jassby et al. 1977b). The high surface-to-volume ratios of most microcosms result in side and bottom effects (e.g., periphyton growth) that exert disproportionately large influences compared to those in natural systems; this problem can be solved by using simple operating procedures (see Harte et al. 1980). Because the small hypolimnion created is quite unrepresentative of natural lakes, realistic conditions of water mixing and thermal stratification are difficult to produce, and in the case of thermal stratification, perhaps undesirable. Additional problems occur because of small populations; thus disproportionate fluctuations in population size may result, masking subtle but important effects.

FIELD ENCLOSURES

The use of field enclosures (e.g., temporary field greenhouses, in terrestrial systems, or corrals in aquatic systems) that partition a representative and manipulatable portion of a natural ecosystem provides a test environment that overcomes many difficulties. The advantage of complete containment of the test system is important for testing the fates and effects of chemicals under field conditions. This approach has been used with some success in marshes (Merks 1968), lakes (Goldman 1962, Hamelink and Waybrant 1973), and forests (Giles 1970, Odum and Jordan 1970). Because controls can be maintained adjacent to the test column, isolating sections of natural environments with plastic film (limnocorrals) has promise for testing chemicals at the level of communities. Because they provide realistic conditions and a broad scope of predictive capability, these types of systems meet the criteria presented earlier. Such enclosures, however, are not easy to maintain or replicate.

FIELD TESTS IN NATURAL ECOSYSTEMS

Tests conducted in natural ecosystems are, of course, the most realistic and provide the best information on ecosystem dynamics. Whether the test is on a small scale, as in a pond or stream, or a large scale, as in a forest or watershed, the unconstrained nature of the system is the defining characteristic. Few extrapolations from species interactions are needed in these tests; systems are selected because they are examples of regions where chemical exposure is expected and because they have characteristics that satisfy the criteria for test systems. Subtle influences and interactions among components of the system that are not present in more simplified laboratory systems can be detected in field experiments (NRC 1975). For example, species at the higher levels of a food chain can be especially sensitive to a chemical, yet these larger organisms are difficult to incorporate into laboratory systems. Therefore, the extent of biomagnification of the chemical for large organisms would be difficult to predict without fully understanding the feeding patterns of the organisms in their natural habitat. Additionally, the mode of entry, insofar as the impact of other contaminants is concerned, will be more realistic if field studies have been conducted to understand the interactions of components within the ecosystem (Cairns and Dickson 1978). Field tests are desirable for enhancing the range of predictability and realism of laboratory data. The validation procedures used in a laboratory are particularly important when attempting to evaluate long-term fates and effects of a chemical.

Despite these important advantages, there are certain limitations inherent to field studies (Cooke 1971, Draggan 1976, Heath 1979, Lighthart and Bond 1976, NRC 1975). Because environmental conditions are not uniform, either in space or in time, field work is difficult to replicate. The ideal isolated environments of the Experimental Lakes

Area of Northwestern Ontario, Canada, and the pondlike terrestrial islands that occur throughout the continent represent virtually identical ecosystems but these are rather rare cases (Johnson and Vallentyne 1971, McCormick et al. 1974). Also, the occurrence of natural variability makes an unequivocal interpretation of field results difficult; results often can be ascribed to variable factors of the system. The spatial heterogeneity of natural ecosystems challenges standardization. The influence of time on environmental variability also is difficult to examine. For instance, a substance may be especially detrimental only during those periods when other environmentally induced stresses (e.g., a prolonged drought) on the test species also are operational.

MODELS

Mathematical models provide a link between actual observations and predictions. If a prediction is similar to an observation but is made for a different place or time, a model can justify the claim that similar properties are involved and similar results are to be expected. Some experimental observations, however, can differ from predictions. For example, a change in mortality observed in the laboratory may suggest alterations of growth rates for the population, although alterations in these rates may not occur in nature. Using observed solubilities of compounds in a model, environmental concentrations can be predicted; but because of many chemical, physical, and biological factors, these levels may not actually occur in the natural ecosystem.

The longer the chain of inferences between actual observation and prediction, the more opportunities there are for error. These include errors in structure of the model as a result of inaccurate assumptions, as well as accumulated errors in data. Models must be tested by using independent estimates of test variables derived from different sources and by conducting analyses of error. Because all models are subject to error despite the use of precautionary measures, inferences gain credibility if they are derived through independent methods of analysis (see Chapter 5 for a more thorough discussion of this point).

STATISTICAL MODELS

Of the many mathematical models in use, statistical models incorporate the fewest theoretical assumptions about the structure and dynamics of the ecosystem. Through measures such as regression coefficients and covariances, techniques of multivariate analysis express associations among the variables of the system. Because these coefficients are properties of a data set rather than characteristics of an ecosystem, statistical models are best used for interpolation rather than extrapolation. These types of models are useful within a narrow range of conditions and can be quite accurate, statistically reliable, and easy to standardize as well as use. They are somewhat unrealistic because no explanation is made regarding operation of the system and because predictions for new situations cannot be made. Such models are formal extensions of a common-sense notion that similar causes have similar results but beg the question, "What makes systems similar?"

SIMULATION MODELS

If a system can be easily described by sets of differential equations derived from established assumptions, then simulation models are most useful. The greatest asset of this type of model is its realism. It is unconstrained by analytical tractability of equations, and its parameters can be selected to represent real properties of a system. For instance, growth rates of a population appear in the statistical model as a regression coefficient of the logarithm of population size against time and in a simulation model as birth rates minus death rates (each measured separately).

Unlike statistical models, errors that are present in simulation models generally result from inaccuracies in the initial design or in estimations of parameters rather than in errors associated with measurement of important variables. The consistency between predictions and observations can be verified with these models but cannot be tested against alternative models.

Simulation models usually are custom designed for a particular use, and are not readily transferable to other computer systems. To date, there has been no attempt to standardize them or to transfer them to new systems. Instead of continuing to generate new models, an effort should be made to apply models already developed to new systems. It would be most useful to develop a generic model that could be used to analyze data from a number of laboratories.

QUALITATIVE AND SEMIQUANTITATIVE MODELS

Models that are qualitative or semiquantitative derive predictions from assumptions about the structure of a system and the network of interacting variables. As long as assumptions about this network are valid, the predictions will be independent, within broad limits, of the precise mathematical forms of interactions and of the magnitudes of effects.

These types of models are tested by using a correspondence between binary (+, −) or trinary (+, o, −) predictions and observations. Error arises from inaccuracy of the underlying assumptions.

If the qualitative variables of structure within these models are similar, conclusions can be drawn for a wide range of systems and tested against alternative assumptions about the network. The models have two major weaknesses: (1) the predictions are not quantitative; and (2) as the number of variables increases, some predictions become ambiguous. The value of these models, however, lies in their strength as indicators of effects and their use to develop explanations of results and indicators of sensitivity by coordination with other models.

INTEGRATED USE OF TEST SYSTEMS

None of the test systems discussed in the above sections satisfactorily meets all the criteria presented at the beginning of the chapter. Each offers certain advantages, and an objective of successful testing strategies should be to exploit the best features of each. An integrated testing approach requires validation of all test procedures and analysis of combined test results. Validation of experimental test procedures involves comparison of data and predictions from one test with those of another, and the final comparison should be made with natural ecosystems. Carefully planned and integrated use of validated test procedures should reduce the need for field test comparisons of effects of chemicals, which would contaminate the natural systems and could be costly and difficult.

Validation of microcosm procedures should be conducted for both natural and disturbed conditions; because of the enormous variety of chemicals that need to be tested, a thorough validation for each cannot be accomplished. A minimally acceptable procedure is to demonstrate that a particular choice of microcosm adequately tracks the parent ecosystem. Major gaps currently exist in the ongoing effort to demonstrate the realism of these test systems, however, and continued research is warranted. Studies that compare microcosm data with information obtained from natural ecosystems are particularly important (see Harris et al. 1980, Harte et al. 1980, Perez et al. 1977 for examples).

Mathematical models should be validated by comparing model predictions with results obtained in laboratory experiments. Such experiments might be conducted using single- or multi-species systems, in laboratories and in the field. Simulation models are particularly valuable for predicting fates of chemicals but require validation at the level of ecosystems. Validation of qualitative mathematical models, which can be of use in predicting effects as well as fates of chemicals, can only be accomplished with test systems that go beyond single-species observations.

Because of the limited application of results from any one test, combinations of data from several types of tests are needed. Those results of simulation models that are of value for predicting the fates of chemicals, for example, can be combined with dose-response information from single-species tests thus providing more complete information about predicted impacts. Similarly, the use of qualitative mathematical analysis of effects to identify points of impact can ease the experimental burden. For example, if models suggest that the chemical will accumulate primarily in aquatic systems, evaluations of effects can be concentrated on aquatic species and test conditions.

Microcosms possess a number of disadvantages, discussed previously, that limit their usefulness when data from these systems are used in isolation from other approaches. The absence of large organisms and the shallow depth of containers make these systems most useful as models of "microvariables" (in aquatic systems, this would include the populations of plankton and microorganisms along with associated chemical flows and system functions). Problems of aging restrict the use of this system to short-term studies. For example, particular species of phytoplankton gradually begin to dominate laboratory microcosms of lake systems, but in the natural environment these species are maintained at relatively low density. Therefore, results from long-term studies using this system may not be representative of responses in the natural system.

A number of ecological properties of concern in ecotoxicology are those that characterize large organisms or large regions of the natural system, and a number of long-term studies are required to assess potential impacts on these macrovariables. By analyzing the effects of chemicals on the smaller organisms and using field data and mathematical models to provide correlations with the large animals or plants, it is possible to obtain information on how both will react to a given environmental contaminant. A recent paper by Gleick (1980) provides a fuller discussion of the point.

Determining the extent of these impacts in microcosms is a complicated problem. The net effect of a disturbance is determined by the magnitude and direction of multi-species responses. Ignoring a single but potentially large response results in misinterpretation of the net direction of the effect. An action that reduces the size of a zooplankton population, for example, may cause a decrease in the population size and productivity levels of those fish that are dependent upon the zooplankton as a food source. The subsequent reduction in grazing by the fish permits an increase in zooplankton populations back to previous levels. This net effect cannot be predicted without some understanding of the relative magnitude of the grazing pressure of fish, the strength of relationships between zooplankton and fish productivity, and numerous other factors that may be important, such as grazing pressures on the fish themselves.

In some cases, long-term effects are best identified by incorporating data from short-term tests into mathematical models. Short-term studies that predict certain chemical impacts on ecosystem properties are ideally suited to laboratory tests. Such impacts include changes in plankton successional patterns or mineralization rates. Long-term evaluations of species interactions are best conducted in the field, despite the many problems associated with field studies. Large field enclosures may be the most useful, although studies that assess the long-term realism of these enclosures are needed. Long-term field monitoring of natural ecosystems that are already polluted could provide valuable insights into chronic impacts, but the opportunities for predictive use of such an approach are quite limited. Much of the current limitation in the capacity for predicting long-term impacts is likely to remain for some time.

Methods for extrapolating data are needed to extend results of tests conducted for a particular system to a wider range of natural conditions. One approach is the development of generic test systems discussed in Hammons (1980). A generic test does not mimic any particular natural system but rather would attempt to represent a wide class of systems. Source material for these tests might be gathered from a variety of natural systems, and the results would be used to characterize or predict the typical behavior of a particular class of ecosystems (e.g., freshwater lakes).

A second approach is to demonstrate the realism of an experimental system by comparing it with a baseline ecosystem from which source materials have been taken. Extrapolation of results from the experimental system to other field situations then can be made, supported by field data and mathematical models. If ecological reserves were used as sources of materials for the experimental systems the second approach would become particularly attractive. (See the discussion of ecological reserves in Chapter 5.) Because at least one natural ecosystem is well simulated, realistic predictions (for that particular parent system) can be expected. Because the parent system would be well characterized through field observations, its relationship to other systems could be relatively well understood, and the problem of extrapolation would be easier.

SUMMARY

Any valid assessment strategy must first establish a well-defined set of criteria for evaluating test procedures. Suitable tests should (a) predict a broad range of phenomena, (b) produce verifiable data, (c) be easily replicable in other laboratories, (d) have a sound statistical basis, and (e) provide data that can be analyzed using accepted statistical techniques. The test systems must be as environmentally realistic as possible in that they duplicate the natural habitat of test species as well as the form and potential fate of the test chemical. The most critical of these criteria are the need to standardize procedures, to provide environmental realism, and to be able to verify test procedures. Meeting these criteria also provides the greatest challenge in test development.

Single-species tests are of considerable value in establishing suitable ranges of dose for use in multi-species tests. These range-finding studies should serve as pilot studies for more integrative test procedures, but must not serve as the prime source of data in an assessment of chemical impacts.

Most test systems currently used do not test population responses to a chemical but only a few individual responses of particular species. More research is needed to develop test procedures that use measurements of changes in population dynamics. High priority should be given to research aimed at making the best possible use of population studies in evaluating effects of chemicals.

Short-term effects of some chemicals can be tested in functional processes of ecosystems (e.g., mineralization or nutrient cycles) using multispecies microcosms. However, the realism of these tests must be verified before they can be used extensively. Long-term effects will be difficult to determine in microcosms without more studies using larger organisms.

An appropriate evaluation should integrate data generated by multiple test systems. No one type of test can provide sufficient information to make accurate predictions of chemical impacts on the environment. Tests should be required to detect effects on a single species, on interactions of species, and on the structure and function of ecosystems. The tests should include laboratory microcosms, mathematical models, and field experiments.

REFERENCES

Alevras, R.A., D.T. Logan, C.B. Dew, E.K. Pikitch, S.S. Moy, R.L. Wyman, J.P. Lawler, S.L. Weiss, R.A. Norris, and W.J. Sydor (1979) Methodology for Assessing Population and Ecosystem Level Effects Related to Intake of Cooling Waters. EPRI-EA-1238. Palo Alto, Calif.: Electric Power Research Institute.

Alevras, R.A., C.B. Dew, E.K. Pikitch, R.L. Wyman, J.P. Lawler, R.A. Norris, and S.L. Weiss (1980) Methodology for Assessing Population and Ecosystem Level Effects Related to Intake of Cooling Waters. Vol. 1: Handbook of Methods Population Level Techniques. EPRI-EA-1402. Palo Alto, Calif.: Electric Power Research Institute.

Cairns, J., Jr. and K.L. Dickson (1978) Field and laboratory protocols for evaluating the effects of chemical substances. J. Test. Eval. 6(2):81-91.

Cooke, A.S. (1971) Selective predation by newts on frog tadpoles treated with DDT. Nature 229:275-276.

Davis, D.E. (1974) Behavior As An Ecological Factor. Vol. 2: Benchmark Papers in Ecology. Stroudsburg, Pa.: Dowden, Hutchinson, and Ross, Inc.

Draggan, S. (1976) The microcosm as a tool for estimation of environmental transport of toxic material. Int. J. Environ. Stud. 10:65-70.

Draggan, S. (1978) TSCA: The U.S. attempt to control toxic chemicals in the environment. Ambio 7(5/6):260-262.

Dudzik, M., J. Harte, A. Jassby, E. Lapan, D. Levy, and J. Rees (1979) Some consideration in the design of aquatic microcosms for plankton studies. Int. J. Environ. Stud. 13:125-130.

Farr, J.A. (1977) Impairment of antipredator behavior in *Palaemonetes pugio* by exposure to sublethal doses of parathion. Trans. Am Fish. Soc. 106:287-290.

Fisher, N.S., E.J. Carpenter, C.C. Remsen, and C.F. Wurster (1974) Effects of PCB on interspecific competition in natural and gnotobiotic phytoplankton communities in continuous and batch cultures. Microb. Ecol. 1:39-50.

Giles, R.H. (1970) The Ecology of a Small Forested Watershed Treated with the Insecticide Malathion-S^{35}. Wildlife Monograph No. 24. Washington, D.C.: The Wildlife Society.

Gleick, P.H. (1980) Lakes and microcosms: Extending microcosm data to aquatic ecosystems. *In* Working Papers for the Committee to Review Methods for Ecotoxicology. Available in limited supply from the Environmental Studies Board, Commission on Natural Resources. Washington, D.C.: National Academy Press.

Goldman, C.R. (1962) A method of studying nutrient limiting factors *in situ* in water columns isolated by polyethylene film. Limnol. Oceanogr. 7:99-101.

Hamelink, J.L. and R.C. Waybrant (1973) Factors Controlling the Dynamics of Non-ionic Synthetic Organic Chemicals in Aquatic Environments. Tech. Report 44. Lafayette, Indiana: Purdue University Water Resources Research Center.

Hammons, A.S. (1980) Methods for Ecological Toxicology: A Critical Review of Laboratory Multi-species Tests. Report No. EPA-560/11-80-026. Oak Ridge, Tenn.: Oak Ridge National Laboratories.

Harris, W.S., B.S. Ausmus, G.K. Eddlemon, S.J. Draggan, J.M. Giddings, D.R. Jackson, R.J. Luxmoore, E.G. O'Neill, R.V. O'Neill, M. Ross-Todd, and P. Van Voris (1980) Microcosms as Potential Screening Tools for Evaluating Transport and Effects of Toxic Substances. Report No. EPA-600/3-80-042. Oak Ridge, Tenn.: Oak Ridge National Laboratories.

Harte, J., D. Levy, J. Rees, and E. Saegebarth (1980) Assessment of Optimum Microcosm Design for Pollution Impact Studies. Final report prepared for Electric Power Research Institute. Palo Alto, Calif.: Electric Power Research Institute.

Heath, R.T. (1979) Holistic study of an aquatic microcosm: Theoretic and practical implications. Int. J. Environ. Stud. 13:87-93.

Jassby, A., M. Dudzik, J. Rees, E. Lapan, D. Levy, and J. Harte (1977a) Production Cycles in Aquatic Microcosms. U.S. Environmental Protection Agency, Office of Research and Development. Report No. EPA-600/7-77-097. Springfield, Va.: National Technical Information Service.

Jassby, A., J. Rees, M. Dudzik, D. Levy, E. Lapan, and J. Harte (1977b) Trophic Structure Modifications by Planktivorous Fish in Aquatic Microcosms. Environmental Protection

Agency, Office of Research and Development. Report No. EPA-600/7-77-096. Springfield, Va.: National Technical Information Service.

Johnson, W.E. and J.R. Vallentyne (1971) Rationale, background and development of experimental lake studies in Northwestern Ontario. J. Fish. Res. Board Can. 28(2):123-128.

Larimore, R.W. and J.M. McNurney (1980) Evaluation of a Cooling-Lake Fishery. Vol. I: Introduction, Water Quality, and Summary. EPRI-EA-1148. Palo Alto, Calif.: Electric Power Research Institute.

Larimore, R.W., J.M. McNurney, and D.R. Halffield, Jr. (1980) Evaluation of a Cooling-Lake Fishery. Vol. 2: Lake Sangchris Ecosystem Modeling. EPRI-EA-1148. Palo Alto, Calif.: Electric Power Research Institute.

Lighthart, G. and H. Bond (1976) Design and preliminary results from a soil/litter microcosm. Int. J. Environ. Stud. 10:51-58.

Logan, D.T., S.S. Moy, R.A. Norris, and W.J. Sydor (1980) Methodology for Assessing Population and Ecosystem Level Effects Related to Intake of Cooling Waters. Vol. 2: Handbook of Methods—Community Analysis Techniques. EPRI-EA-1402. Palo Alto, Calif.: Electric Power Research Institute.

May, R.M. (1976) Theoretical Ecology: Principles and Applications. Philadelphia, Pa.: W.B. Saunders Co.

McCormick, J.F., A.E. Lugo, and R.R. Sharitz (1974) Experimental analysis of ecosystems. Pages 151-180, Handbook of Vegetation Science. Part VI: Vegetation and Environment, edited by B.R. Strain and W.D. Billings. The Hague, The Netherlands: Dr. W. Junk B.V.

Merks, R.L. (1968) The accumulation of ^{36}CL ring-labelled DDT in a freshwater marsh. J. Wildl. Manage. 32(2):376-398.

Mosser, J.L., N.S. Fisher, and C.F. Wurster (1972) Polychlorinated biphenyls and DDT alter species composition in mixed cultures of algae. Science 176:533-535.

National Research Council (1975) Appendix D. Environmental toxicology: Supplementary material to chapter XIII. *In* Principles for Evaluating Chemicals in the Environment. Environmental Studies Board and Committee on Toxicology. Washington, D.C.: National Academy of Sciences.

Odum, H.T. and C.F. Jordan (1970) Metabolism and evapotranspiration of the lower forest in a giant plastic cylinder. Pages 165-190, A tropical Rainforest, edited by H.T. Odum. Washington, D.C.: Division of Technical Information, U.S. Atomic Energy Commission.

Perez, K.T., G.M. Morrison, N.F. Lockie, C.A. Oviatt, S.W. Nixon, B.A. Buckley, and J.F. Heltsche (1977) The importance of physical and biotic scaling to the experimental simulation of a coastal marine ecosystem. Helgolander wiss, Meereunters 30:144-162.

Sharitz, R.R. and J.F. McCormick (1973) Population dynamics of two competing annual plant species. Ecology 54:723-740.

Whittaker, R.H. (1961) Experiments with radio-phosphorous tracer in aquarium microcosms. Ecol. Monog. 31:157-188.

Zaret, T.M. (1972) Predators, invisible prey, and the nature of polymorphism in the Cladocera (class crustacea). Limnol. Oceanog. 17:171-184.

Zaret, T.M. (1975) Strategies for existence of zooplankton prey in homogeneous environments. Verh. Int. Verein. Limnol. 19:1484-1489.

Zaret, T.M. and W.C. Kerfoot (1975) Fish predation on *Bosmina bongirostris*: Body-size selection versus visibility selection. Ecology 56:232-237.

5 Assessment Strategy

Previous chapters have discussed the need to determine impacts of chemicals using both single-species and ecosystem tests. Single-species tests alone cannot provide sufficient data with which to predict damage to an entire ecosystem. Although results of such tests may be quite accurate for laboratory populations, they do not provide an accurate estimate of either a species response in the natural environment or indirect effects that might occur in other components of a system. The interrelationships among species, the continuity of functional processes, and the evolutionary history of a system are all factors that determine the response of an ecosystem to pollutants.

Since the passage of TSCA, government officials and environmental scientists have developed new strategies for a logical progression through specific testing schemes. The EPA has published proposed guidelines for testing procedures that suggest a tiered approach to generating data needed for estimating potential environmental effects of new chemicals (U.S. EPA 1979). Publications of the American Society for Testing and Materials and the American Fisheries Society suggest variations of a tiered approach for an assessment strategy (Cairns et al. 1978, Dickson et al. 1979). For the most part, these guidelines and testing strategies rely heavily on single-species tests that use a standard set of laboratory populations (e.g., *Daphnia*, rainbow trout, rabbit, and mouse). These strategies progress in steps, starting with the collection of data on chemical and physical properties of the test substance, proceeding to data on acute toxicity, and usually ending with data on chronic toxicity. Data on bioconcentration or biomagnification and ecosystem interactions are the last to be collected, if obtained at all. Generally, movement through each tier is dependent on the results from a previous level. Chronic toxicity tests, for example, are conducted only if data indicate acute toxicity for the test population. This chapter suggests a different approach for evaluating effects of chemicals on ecosystems.

DESIGN FOR ESTIMATING ENVIRONMENTAL EFFECTS

It is difficult for EPA to carry out the provisions of TSCA given current approaches to evaluating effects of chemicals. The recommended strategy discussed in the following pages should provide the type of information and data that will support scientifically based predictions about potential hazards. The strategy includes establishment of baseline ecosystems, use of multilevel integrated tests, evaluation of interactive impacts, and a method to reduce uncertainty of predictions.

BASELINE ECOSYSTEMS

The ability to predict the environmental impact of a substance requires an understanding of ecosystem structure and function, and baseline studies of ecosystems are needed to provide the understanding. The studies should include: (1) field observations in selected ecosystems, (2) experimental work directed toward understanding the dynamics of the ecosystems, and (3) mathematical models to facilitate understanding the interactions of components within the system.

Ecological reserves could be established throughout the country as sites for the field observations of baseline ecosystems. The concept of biological reserves is not new. The Man and the Biosphere Program of UNESCO established 27 sites in the United States (Franklin 1977). Project AQUA of the International Biological Program established a large number of reserves throughout the world, many of them within the United States (Luther and Rzoska 1971). Selected ecosystems representative of geographic regions and unique or fragile systems should be studied and the results used in identifying key species, properties of populations, and ecosystem processes. The knowledge acquired can serve as a starting point in choosing subsequent test approaches.

Laboratory work will be needed to understand the dynamics of changes caused naturally in the systems. Results of field observations and laboratory data should be combined to develop general models for each system. The models can then be used to predict changes in the ecosystem that may result from the introduction of chemicals.

MULTILEVEL INTEGRATED TESTING

Integrated testing (as discussed in Chapter 4) is proposed as an alternative to the tiered or hierarchical testing currently practiced. The recommendation is based on two considerations: (1) with the exception of compounds that are biologically inactive, valid inferences about changes in ecosystem phenomena cannot be made from observations limited to lower levels of biological organization; (2) the results of one type of test can determine further research using another test system (e.g., microcosm test results may suggest the need for tests of effects on a particularly sensitive species).

An integrated testing approach should incorporate four classes of information. Data must be obtained on:

1. characterizations of the test chemical, including chemical and physical data, estimates of fates in and among ecosystems, and estimates of environmental concentrations and exposure time;

2. physiological responses of species indicating morphological, biochemical, genetic, and pathological changes related to the presence of a chemical;

3. changes in species interactions (e.g., predation, competition, and migratory behavior) noting changes in sizes of populations or structure of the ecosystem; and

4. changes in the functional processes of the ecosystem (e.g., nutrient flows and mineralization).

If the ecosystem of concern is already under stress, the impact of a chemical can be greater. Because ecosystems are increasingly subject to many kinds of stresses as a result of human activity and natural phenomena, a realistic program for evaluating impacts should ask two questions: (1) Does the chemical have a greater impact in the presence of other contaminants, or during extreme climatic conditions? (2) Does the chemical increase the vulnerability of components of the ecosystem to other stresses, such as disease, parasitism, or predation? Information should be acquired from both single-species and experimental ecosystem tests done under "standard" stress conditions, and the standard set of stressors should be developed for each type of ecosystem. Stresses might include: (a) stresses on primary producers (e.g., nutrient scarcity, reduced photosynthesis, water-short plants); (b) stresses on consumers; (c) demographic stresses (e.g., reduction or augmentation at lower trophic levels, removal or addition of top consumers, dilution and concentration of the whole system); (d) pervasive stresses such as temperature or moisture variations; and (e) specialized stresses that affect behavior of species. The specific list of stress conditions should be determined for each kind of ecosystem, with emphasis on those that are most likely to maximize the impact of chemicals. In some instances, tests can be conducted simultaneously; but in others, they must be done sequentially (e.g., data from [1] above will be needed before appropriate species and ecosystem processes can be chosen).

The current state-of-the-art of ecotoxicology does not enable us to make statements regarding specific test procedures. At this point it is most prudent to identify the types of data needed and the suitable test systems (as discussed in Chapter 4) that can be used to obtain the data. Thus as knowledge in this field increases and more experience is gained, particular test procedures and species can be modified. It must be emphasized, however, that decisions about the use of a chemical and the need for its regulation must not be made until all classes of information are available.

REDUNDANCY IN THE FACE OF UNCERTAINTY

Only direct observation after release of a substance to the environment can provide error-free information on the substance's impact, but direct observation is precisely what we are trying to prevent. Thus reliance on test methods that are beset with margins of error becomes necessary. Because different methods are likely to suffer from different types of error, a decision about the use of a chemical should be based on the convergence of different lines of evidence, each with different combinations of observations and inferences. In order to have confidence in the conclusions of an evaluation process, it is necessary to do the following:

(1) understand the pathways of the chemical into the environment from the initial source to the final place of degradation or accumulation, and identify those ecosystems at potentially greatest risk (e.g., soil, lakes, or forest);

(2) understand the structure and function of at least three sample ecosystems to know what properties might be most affected;

(3) identify potential points of biological impact for populations within sample ecosystems as well as single- and multi-species interactions;

(4)   demonstrate in laboratory test systems that the anticipated impact does occur;

(5)   verify the significance of the impacts in light of established models;

(6)   evaluate the influence that natural selection may have on the response of the ecosystem.

IMPLEMENTATION OF THE ASSESSMENT STRATEGY

Before the assessment strategy can be implemented, several preparatory steps must be taken. The initial phase of selecting sites and standardizing measures, stressors, and perturbations could be accomplished in 1 to 2 years and the system could be operative in 5 to 10 years.

(1) *Selection of the baseline ecosystems.* Baseline systems should be chosen with the intent of including major and especially vulnerable ecosystems in different geographic regions. A variety of ecosystems in a number of locations, both terrestrial and aquatic, should be included in the baseline systems. Those with existing substantive data bases should be given reference. Natural systems (ecological reserves), economically exploited systems (agricultural, forest, and aquatic), and recreational areas should be represented. Recent conferences held by the National Science Foundation provide details about how reserves might be established (Botkin 1977, 1978; The Institute of Ecology 1979). Existing field research and agricultural experiment stations as well as established biosphere reserves could serve as nuclei for establishing baseline systems. Initially, 20 to 30 sites may be needed. As experience and knowledge about ecosystem structures and functions are gained, a reduction in number may be justified.

(2) *Characterization of each baseline ecosystem.* Taxonomic surveys, estimates of population size, and data on system dynamics should be obtained. A minimal list of measurements and the conditions under which they are to be taken must be standardized. At least four sets of seasonal observations are needed, but the precise times should be determined according to climatic events (e.g., lake turn-over or snow melt) rather than by calendar dates. Characteristics to be measured should include (a) the degree of diversity and a list of components, (b) biomass and productivity, (c) nutrient and energy flows (e.g., mineralizaton and nitrogen fixation) and (d) population dynamics and physiological states of selected species. Species should be selected on the basis of direct interest (e.g., endangered species), their role in a recognized system function (e.g., erosion control, detoxification or mineralization), eco-

nomic importance, or their role in the maintenance of the ecosystem. Evaluations of the physiological state should include species tolerance to a standard set of stress factors.

(3) *Establishment of experimental ecosystems.* A monitoring and experimental unit should be associated with each baseline study. Test material for the laboratory systems should be taken from the baseline system. Experimental ecosystems would represent the baseline reserve if: (a) the measurements track those of the baseline system or; (b) in those cases where the experimental unit lacks major components of the parent system, the results of simulated impacts track the parent system (e.g., removal of zooplankton from the aquarium at a rate proportional to fish predation in a lake). Procedures for testing the selective impact of changes in mortality, fecundity, growth rates, and predation rates on species in these experimental units should be developed.

(4) *Development of a model of the dynamics of the baseline ecosystem.* Models of the dynamics involved, based upon direct experimentation and observation, should be developed that predict limits of resilience and sensitivity. When a set of mathematical models has been developed that adequately tracks the baseline ecosystem, the disturbed experimental system, and changes in the experimental system after a specific set of perturbations, the set will be accepted as representing an ecosystem.

Once these preparatory steps have been accomplished, it will be possible to begin testing the effects of chemicals. The following scheme should be completed.

1. The chemical and physical properties of the substance together with the circumstances of its production and use should be used to determine which ecosystems may be the direct recipients of the chemical.

2. For the ecosystems in question, the key species and processes, as identified in the baseline ecosystem, should be used in single-species and microcosm tests. The tests will be used to determine potential points of entry and direct impact.

3. With data obtained from these tests, the set of models is then used to predict responses, direct and indirect, of the ecosystem.

4. The model predictions should be then checked experimentally by manipulating the experimental (laboratory) baseline system to simulate the impact of the chemical.

5. The final step is to test for anticipated effects of the chemical using a multispecies integrated approach.

The information obtained in 3, 4, and 5 must be in agreement before decisions on the potential hazard can be made. If there is a discrepancy between 3 and 4, it could mean either that the model is flawed or that the manipulations in 4 have effects other than those intended. This implies error in designing the experimental unit. If results of 3 and 4 are similar but different from 5, it would suggest that the chemical has impacts other than those previously identified and that further work is needed. When 2, 3, 4, and 5 agree, the likely effect of the chemical on the ecosystem will be known.

Finally the possibility exists that some species may respond to the chemical by altering rates of mortality, fecundity, or interspecific interactions. The models should indicate where such changes may have important effects. Further testing to determine genetic variance and adaptive responses would then be justified.

The purpose of an evaluation strategy is to protect ecosystems. As the number of chemicals used by society increases, the probability of serious problems will also increase. The proposed assessment strategy is aimed at detecting even impacts of low probability by combining results of different types of tests conducted under different circumstances. As experience accumulates, however, strong correlations among tests may emerge that allow reduction in the number of tests performed. Therefore the responsible regulatory agency should periodically review accumulated experience to determine whether elimination of some tests might be justified. New understanding could also lead to requirements of additional observations. Ideally, the evaluation process should evolve toward one in which more information is obtained from fewer experiments.

Implementation of the assessment strategy will require more individuals trained in the field of ecotoxicology. More funds should be made available to meet this need.

VALIDATION OF MODELS USED IN THE TEST STRATEGY

The end point of an assessment strategy is the claim that a given substance is likely (or unlikely) to have an adverse effect on either general or specific environmental properties. As discussed previously, the evidence for that claim will necessarily be indirect, because the intent of the assessment is to *prevent* the predicted impact. Thus conclusions are drawn on the basis of theoretical models and on direct observation using laboratory or field experimentation. The reliability of any strategy depends on an accurate prediction of the fate of a chemical from its point of release to points of biological impact and subsequent movement to other systems. The following approach can be used to validate the modeling process.

(1) *Models of the natural system.* Models can be developed to illustrate the behavior of natural systems and can be either qualitative or quantitative. Quantitative models are difficult to develop, however, because of their dependence on a large quantity of data and a sensitivity of the outcomes to the functional forms of equations as well as to the variation in parameters. Once a simulation model has been developed and verified, it can give short-term predictions about the system from which it was derived; the capability to generalize to other systems is questionable.

Qualitative models examine the structure of a network of interacting variables within a system. These models are often insensitive to many details that are important in simulation models, and the resulting predictions can be only qualitative (i.e., indicating two $(+, -)$ or three $(+, o, -)$ alternatives). Therefore, each prediction is only a weak confirmation of the model, but the results of several models can be combined to strengthen the confirmation.

(2) *Comparison of different models.* Because one model (even if considered as a "best-guess model") might omit important relationships, confidence in a prediction is enhanced if alternative models yield similar results. If dissimilar models predict different points of impact or differences in the quality of an impact, however, then further investigation is needed.

(3) *Experimental results.* Data from multilevel test conditions (natural and disturbed) should be obtained to track the behavior of an ecosystem. Any deviations of the experimental results from those predicted by models (1 above) suggest that further work is needed to characterize the system being studied.

(4) *Correlations between experimental and model predictions.* A correlation between experimental response and model prediction also will strengthen the estimates of identified points of impact. If results from tests identifying a point of impact are incompatible with results from tests identifying biological responses, the models are not reliable reflections of the ecosystem being investigated.

Validation of a model and its resulting predictions should be independent of the type of substance being tested. A model must define interactions within a system in such a manner that points of entry and impact can be predicted for *any* chemical, provided that data on chemical and physical properties of the compound are known. Although many tests may be required to validate a model, once it is established as being generically precise, confirmation will not be needed for each new class of chemicals.

A model must include general information about the composition and interactions of a particular ecosystem and specific information about the test substance. The concentration and form of the substance at the point of impact, the nature of the impact, and the transfer coefficients throughout the trophic structure are very important factors. The following considerations should guide the development of generic models.

• Concentration and distribution of chemicals within the natural environment can be extraordinarily variable; thus, at some place and time, local concentrations can be several orders of magnitude higher (or lower) than a predicted average. A model of the system should consider the consequences of such extremes.

• The transfer of chemicals among the biota also can vary and will depend upon such factors as rates of feeding, patterns of behavior, and alternative food sources at different stages in the life cycle of species.

• Some compounds may be transformed as they move through the system, and the degree of toxicity of any transformation products may be different from that of the parent compound. For example, consider the following simple model illustrating interactions between a nutrient source (N), a primary producer (P) and a consumer (C):

In the laboratory the test substance causes increased mortality for P but does not appear to affect C. The predicted outcome is that the population size of P remains stable and that the population size of C declines. The population size of P is unchanged because increased mortality caused by the substance would be balanced by a decrease in predation by C; thus, the impact is absorbed by C. If the compound is transformed by P to a product that is lethal to C, however, the increased mortality of the predator reduces predation pressures on the prey. The result may be an increase in the population size of P with subsequent heavy grazing on N. Therefore the flow from N to P, and hence the productivity per unit biomass of P, is reduced. Because per capita consumption of P increases, C has higher birth and death rates and a younger population, while P has lower birth and death rates and, therefore, an older population.

If these unexpected results appear first in microcosm experiments, they might be interpreted as a direct external impact on C. This result can be accepted as evidence of an unexpected physiological impact, and the model and resulting predictions must be altered. Any other species connected to this subsystem through consumption of N would respond as if the population size of N were reduced.

The system-level evidence for transformation of the test substance, however, would be strengthened by the direct demonstration that when C feeds on P—contaminated with the test substance—the organisms die. But no effect occurs when C is fed on gelatin pellets containing the test substance.

This type of experiment can be done either as a matter of course in the assessment scheme or only if results of the microcosm experiments indicate possible transformation. The choice depends on whether this effect is expected based on previous experience or whether the willingness exists to use independent lines of evidence in confirming a prediction.

• If a system component (e.g., large predators) cannot be included in a microcosm experiment, the position of that component in the ecosystem must be verified using models that estimate the transfer coefficients among trophic levels and the impact of contaminated prey and that simulate the role of predators by some pseudopredation mechanism (e.g., sieving out the prey).

If predictions have been made and confirmed using the four-part approach described previously, results can be extrapolated to other, similar ecosystems. Because of the following points, caution must be used in the interpretation of data and extrapolation of results.

1. The new system may have the same structure as the previous system, but differences may exist in the magnitude of parameter responses; thus, the extent of these differences must be determined.

2. Despite similar structures, ecosystems may have different points of impact due to species-specific sensitivities; the importance of these differences must be determined.

3. Qualitative structural differences might exist between two systems; the extent of these differences might not be quantifiable.

CONSIDERATIONS OF VARIABILITY

In the best-designed experimental studies, data are variable, i.e., repeated measurements of any given phenomenon are not identical. This point is of particular importance in ecotoxicology. The numerous steps in data collection use a variety of biotic and abiotic measurements with typically wide ranges of variance. Additionally, there is variability in the properties of test chemicals, as discussed previously (see Variables Affecting Fates, Chapter 2). The variance in ecosystem response also can be large because of differences in species composition, the occurrence of geographic races, differences in life-cycle stages, age structure, sex, vigor of the population, as well as the presence of other materials that enhance or counteract biological effects of the test chemical. Not only do individuals and populations differ in responses to the presence of a given substance, but there is variability among system responses, even for very similar ecosystems.

This inescapable variability means that any measurement indicating a change in some property or condition will be imprecise; thus any conclusion based on the data also will have some inherent errors. Statisticians reduce the imprecision by estimating the error, expressed as a confidence interval of the conclusion. Scientists approach the problem by depending not on one test procedure only, but on a number of different procedures leading to a common conclusion with enhanced credibility. In addition, sampling techniques are used in such a way that estimates of major sources of variability are provided. The use of either a single measure or a mean value for a phenomenon without also expressing a quantitative estimate of variability is inappropriate.

Because the sum of these measures is most important, data from two or more measures are added in some assessments. Each original quantity, however, has an error associated with it, which is expressed as a variance (e.g., the square of the standard error). In statistical terms, the variance of a sum of values equals the sum of the variance of individual values, and the magnitude of error in these summations is very important.

As a hypothetical example, consider that a productivity measure (P) is determined as $P = A + B - C$, where A, B, and C are individually measured values, each having an error associated with it.

measure	value	variance	standard error	confidence interval
A	10.0	−4.00	2.0	10.0 ± 4.0
B	25.0	−8.00	2.8	25.0 ± 5.6
C	20.0	10.00	3.2	20.0 ± 6.4

From these values the final measure of P is $10.0 + 25.0 - 20.0 = 15.0$, and the variance of P equals $4.00 + 8.00 + 10.00 = 22.00$; the standard error is 4.7, and the confidence interval is approximately 15.0 ± 9.4 (with the exact interval dependent on the sample size). (Note that variances of negative terms in a sum are added, just as variances of positive terms are.)

Sometimes a final data point is a product of more than one term. If logarithms were used, this situation could be handled as in the above example, (i.e., if $W = XY/Z$, then $\ln W = \ln X + \ln Y - \ln Z$, and the variance of $\ln W$ = variance of $\ln X$ + variance of $\ln Y$ + variance of $\ln Z$). An approximate confidence interval for ln W can be obtained as $\ln W \pm 2(\text{variance of } \ln W)^{1/2}$. Thus the antilogarithm of $\ln W - 2(\text{variance of } \ln W)^{1/2}$ represents approximately the lower confidence limit for W, and the antilogarithm of $\ln W + 2)\text{variance of } \ln W)^{1/2}$ is an approximation of the upper confidence limit for W.

This last procedure has been useful in applying various correction or conversion factors in toxicological studies (U.S. EPA 1978). Because of a lack of data for a wide range of species, development of some correction factor has been necessary. For example, both 96-hour and 48-hour LC_{50} values can be determined for individuals of a particular test population (e.g., rainbow trout). The geometric mean of the ratio of the 96-hour value to the 48-hour value might be 0.85 with a variance of 0.1812 (the logarithms of the ratios). Additional measurements from a 48-hour study are converted to a predicted 96-hour value simply by multiplying the data by 0.85. The variance associated with this predicted value is the variance of the 48-hour value + 0.1812 (the latter being the variance of the 0.85 conversion factor). These types of conversions are common practice. In series of two to four conversions (and these are common), the error of the prediction is compounded.

For precise and accurate predictions, conversion factors are best applied to homogeneous groups. In the above example, fish of the same species, sex, and age *must* be used. These ideal factors are not always available, however, and investigators have resorted to using factors determined for heterogeneous populations; at times predictions for saltwater populations are made using data obtained with freshwater organisms, or invertebrate data are used to predict values for fish.

An acceptable measure of variability is obtained by replicate sampling (i.e., collecting more than one data point for each quantity studied). In general, the larger the sample size (e.g., the greater the numer of replicates obtained), the smaller the variance and error of prediction with the result being greater precision. Obtaining only a few replicates results in a useless, or at best inconclusive, study because of the low confidence that can be placed in the results. It should be noted, however, that a point will be reached at which increased numbers of replicates obtained no longer repay the increased costs of time, effort, and material.

Another important reason for replicate sampling is the need to estimate extreme values. Extreme values are observed infrequently, but
when they occur, their effect can be considerable. In many cases, extremely high (or, conceivably, extremely low) values of a particular
measure are more important for predicting the impact of a chemical
than are mean values.

SUMMARY

To determine the impact that a chemical could have on an ecosystem,
information is needed about the system's natural or undisturbed state.
A properly designed assessment strategy must include, in addition to
an experimental situation, a data base reflecting natural conditions.

An evaluation process should include use of baseline ecosystems, a
multilevel integrated testing scheme, and verified mathematical models
of ecosystems. Four essential classes of information can be obtained
using the integrated testing scheme: characterization of the test chemical, physiological responses of individuals to the presence of a chemical,
changes in interactions of species, and changes in functional processes
of ecosystems. Testing for effects of chemicals should include conditions
of stress to determine the influence of stress on the final impact of a
chemical.

The assessment strategy can be implemented while obtaining a thorough understanding of the natural environment, and an approach to
achieving this understanding is provided. In addition, a five-step approach is recommended for testing for impacts of chemicals. The approach includes identification of chemical and physical properties of the
test substance, determination of potential points of entry into and impact
upon the ecosystem, predictions of direct and indirect responses using
mathematical models of the system, verification of the predictions of
the model by manipulations of laboratory representations of baseline
systems, and substantiation of anticipated effects using a multilevel integrated testing scheme.

Verification of a general model of an ecosystem can be made through
the use of several types of models of the natural system, comparison of
results from different models, generation of data in experimental systems
for disturbed and undisturbed ecosystems, and correlation of experimentally derived estimates of impact and predictions based on different
models.

Because of the properties of chemicals and ecosystems, variability in
data is inevitable. Precaution must be taken to reduce the variance as
much as possible. Calculating confidence limits on data can express

uncertainties about the precision of predictions. Correction or conversion factors are useful only if applied to homogeneous test conditions and populations. Replicate sampling is perhaps the most acceptable way to determine variance and error associated with predictions.

An integrated evaluation process as described in this chapter should be developed to predict environmental impacts of a chemical. The strategy should allow detection even of impacts of low probability by combining several types of data collected under different conditions. Indication of an adverse impact from any group or class of information should be sufficient cause for concern and further testing, but a conclusion about the impact of a chemical and the need for regulating its use would require consistent indications from all sources of data.

Because of the lack of individuals trained in ecotoxicology, implementation of the assessment strategy will be difficult. To remedy this situation, more funds should be made available to support the training of students in ecotoxicology.

REFERENCES

Botkin, D.B., ed. (1977) Long-term Ecological Measurements. Washington, D.C.: National Science Foundation.

Botkin, D.B., ed. (1978) A Pilot Program for Long-term Observation and Study of Ecosystems in the United States. Washington, D.C.: National Science Foundation.

Cairns, J., Jr., K.L. Dickson, and A.W. Maki, eds. (1978) Estimating the Hazard of Chemical Substances to Aquatic Life. Special Technical Publication 657. Philadelphia, Pa.: American Society for Testing and Materials.

Dickson, K.L., A.W. Maki, and J. Cairns, Jr., eds. (1979) Analyzing the Hazard Evaluation Process. Washington, D.C.: American Fisheries Society.

Franklin, J.F. (1977) The biosphere reserve program in the United States. Science 195:262-267.

Luther, H. and J. Rzoska (1971) IBP Handbook #21. Oxford and Edenburg: Blackwell Scientific.

The Institute of Ecology (1979) Long-term Ecological Research: Concept Statement and Management Needs. Washington, D.C.: National Science Foundation.

U.S. Environmental Protection Agency (1978) Water quality criteria. Federal Register 43(97):21506-21518.

U.S. Environmental Protection Agency (1979) Toxic substance control: Discussion of premanufacture testing policy and technical issues. Federal Register 44(53):16240-16292.

Environmental Partitioning

This discussion of environmental partitioning is taken directly from McCall et al. (1980), a paper commissioned by the NRC Committee to Review Methods for Ecotoxicology.

As noted in Chapter 2, a number of factors are important in determining the fate of chemicals in aquatic and terrestrial ecosystems. Understanding the complex process of entry, movement, and disposition of chemicals in any given system can be enhanced by consideration of the chemical and physical phenomena associated with interrelationships among the environmental media and between these media and the chemical itself. Association with biota is pertinent to both interactions. Because the same partitioning processes occur in all ecosystems regardless of type, it is only the size, type, and number of compartments in a given system that will determine the distribution pattern of a chemical in that system. Selected interactions are discussed below. The reader should refer to Chapter 2 as needed.

SOIL/WATER INTERACTIONS

The capacity of soil to adsorb a chemical in solution can be described by the Freundlick equation, $x/m = KC^{1/n}$, where

x/m = amount (x) absorbed per unit amount of adsorbent (m)
K = equilibrium constant
C = equilibrium concentration
n = the degree of nonlinearity

Or, more generally:

$$Kd = \frac{\mu g \text{ chemical/g soil}}{\mu g \text{ chemical/g water}} \qquad (1)$$

where

> Kd = sorption coefficient
> μg chemical/g soil = concentration of adsorbed chemical
> μg chemical/g water = concentration of chemical in solution

The sorptive capacity of soil is directly proportional to the level of organic carbon (i.e., soils with high levels of organic carbon have high sorptive capacities). Thus, the sorptive characteristic of a chemical can be normalized to derive a sorptive constant (K_{oc}) based on organic carbon content, independent of other soil characteristics.

$$K_{oc} = \frac{\mu g \text{ chemical/g organic carbon}}{\mu g \text{ chemical/g water}} \qquad (2)$$

Therefore

$$K_{oc} = \frac{Kd}{\% \text{ organic carbon}} \cdot 100\% \qquad (3)$$

WATER/AIR INTERACTIONS

Henry's Law describes the distribution of a chemical between water and air. The expression can be written as

$$H = \frac{C_{air}}{C_{water}} = \frac{PM}{RT(WS)} \qquad (4)$$

where

> H = Henry's Law Constant = $1/K_w$
> C_{air} = concentration of the chemical in air (mg/l)
> C_{water} = concentration of the chemical in water (mg/l)
> P = vapor pressure of pure chemical (mm Hg)
> M = molecular weight
> T = temperature (°K)

R = 0.08205 1-atm/deg-mole

WS = water solubility (ppm)

Equation (4) can be simplified and expressed in terms of K_w, the reciprocal of H.

$$K_w = \frac{T(WS)}{16.04\ PM} \tag{5}$$

CHEMICAL/ORGANISM INTERACTIONS

A bioconcentration factor (BCF) is frequently used to describe the partitioning of a chemical between an organism and water (aquatic systems) and between an organism and its diet (terrestrial systems). Its use in aquatic systems is probably the more common. In fish, for example,

$$BCF = \frac{\mu g\ chemical/g\ fish}{\mu g\ chemical/g\ water} \tag{6}$$

where

BCF = bioconcentration factor

μg chemical/g fish = concentration of chemical in fish

μg chemical/g water = concentration of chemical in water

MODEL ECOSYSTEM INTERACTIONS

This example illustrates the use of the individual partition coefficients discussed above (K_{oc}, K_w, and BCF) in estimating the distribution of a chemical in a time-dependent ecosystem model. The coefficients are derived from equilibrium-based models.

Consider the model ecosystem depicted in Figure A.1. Dimensions of the system are presented in Table A.1. In addition, assume that the surface is 30 percent water with a 10 m deep pond, and that the equilibrium process is limited to the top 7.5 cm of soil and 5 cm of sediment. The overall equilibrium expression for the entire system can be represented as follows:

$$C_{sed} \underset{}{\overset{Kd_{sed}}{\rightleftharpoons}} C_w \underset{}{\overset{K_w}{\rightleftharpoons}} C_a \underset{}{\overset{K_w}{\rightleftharpoons}} C_{sw} \underset{}{\overset{Kd_{soil}}{\rightleftharpoons}} C_s \tag{7}$$

$$\Big\Updownarrow BCF$$

$$C_f$$

where

C_{sed} = weight of chemical in sediment
C_w = weight of chemical in water
C_f = weight of chemical in fish
C_a = weight of chemical in air
C_{sw} = weight of chemical in soil water
C_s = weight of chemical in soil

The atmosphere is the principal compartment in which transfer of the chemical between aquatic and terrestrial segments of the ecosystem occurs. The air is considered to be in equilibrium with soil water (which is assumed to be 25 percent of the soil compartment) and water in the aquatic segment.

Assuming a given volume percentage for each compartment, such that

$$\% \text{ air} + \% \text{ water} + \% \text{ sediment} + \% \text{ fish}$$
$$+ \% \text{ soil water} + \% \text{ soil} = 100\% \quad (8)$$

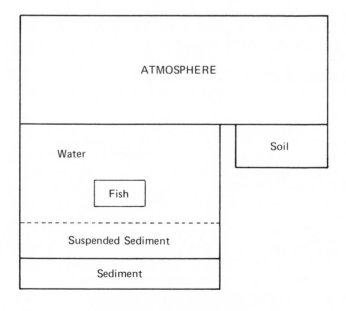

FIGURE A.1   Model ecosystem.

TABLE A.1   Characteristics of Model Ecosystem in Figure A.1

		Volume
Atmosphere	-1000 m $\times$ 1000 m $\times$ 10 km	$= 10^{10}$ m^3
Water	-1000 m $\times$ 300 m $\times$ 10 m	$= 3 \times 10^6$ m^3
Soil[a]	-1000 m $\times$ 700 m $\times$ .076 m	$= 5.4 \times 10^4$ m^3
Sediment	-1000 m $\times$ 300 m $\times$ .05 m	$= 1.5 \times 10^4$ m^3
Suspended sediment	$- \sim 10$ ppm in water	$= 15$ m^3
Fish	-1 ppm in water	$= 3$ m^3
Soil organic carbon		$= 2\%$
Sediment organic carbon		$= 8\%$

[a] The total volume of soil is assumed to be composed of 25% air, 25% water, and 50% soil solids.

the partition expressions must be written in terms of volumes by considering densities (p) of the media. Soil and sediment are assumed to have a density of 2.5 g/cc, and water and fish, 1 g/cc.

For the ecosystem then,

$$\%C_{sed} + \%C_w + \%C_f + \%C_a + \%C_{sw} + \%C_s = 100\% \quad (9)$$

The sum of the percentage of chemical in all compartments equals 100 percent. The partition expressions then become

$$\text{sediment} \rightleftharpoons \text{water: } Kd_{(sed)} = \frac{\%C_{sed}/(\% \text{ sediment}) \, 2.5}{\%C_w/\% \text{ water}} \quad (10)$$

$$\text{water} \rightleftharpoons \text{fish: } BCF = \frac{\%C_f/\% \text{ fish}}{\%C_w/\% \text{ water}} \quad (11)$$

$$\text{water} \rightleftharpoons \text{air: } K_w = \frac{\%C_w/\% \text{ water}}{\%C_a/\% \text{ air}} \quad (12)$$

$$\text{soil water} \rightleftharpoons \text{air: } K_w = \frac{\%C_{sw}/\% \text{ soil water}}{\%C_a/\% \text{ air}} \quad (13)$$

$$\text{soil water} \rightleftharpoons \text{soil: } K_w = \frac{\%C_s/(\% \text{ soil}) \, 2.5}{\%C_{sw}/\% \text{ soil water}} \quad (14)$$

These expressions can be combined to show that once the percentage of chemical in the water has been calculated, that value can then be used to derive the percentage of chemical in each of the other compartments. For example,

$$\%C_w = 100 \div \left[1 + Kd_{(sed]}\left(\frac{\%\text{ sediment}}{\%\text{ water}}\right)2.5 + BCF\left(\frac{\%\text{ fish}}{\%\text{ water}}\right) \right.$$

$$\left. + 1/K_w\left(\frac{\%\text{ air}}{\%\text{ water}}\right) + \frac{\%\text{ soil water}}{\%\text{ water}} + Kd_{(s)}\left(\frac{\%\text{ soil}}{\%\text{ water}}\right)2.5 \right] \quad (15)$$

REFERENCE

McCall, P.J., D.A. Laskowski, R.L. Swann, and H.J. Dishburger (1980) Partitioning of Chemicals in Model Ecosystems. Commissioned paper prepared for the Committee to Review Methods for Ecotoxicology, Environmental Studies Board, Commission on Natural Resources, National Research Council. (Unpublished)

Alternative Models for Evaluating Connectivity

When the interrelationships among components of an ecosystem are not known, impacts of a chemical cannot be predicted unless alternative models are used. Figure B.1 illustrates 15 alternatives to the model shown in Figure 3.1 (see Chapter 3); a number of variations of the interactions among a nutrients source (N), two consumers (H_1, H_2), and a predator (P) are shown. For each model, a matrix indicates the direction of change for the equilibrium level or the average value of the variable (e.g., component) listed above each column. The directed impact of a chemical enters the system as a positive input through the variable at the left of each row (see Figure B.1).

The analysis shows that for many of the predictions the detailed structure of the system does not matter. The predictions that coincide under different models can be characterized as strong. Those predictions that differ from the first model are enclosed in boxes and permit differentiation among models if more observations are made. When predictions do not coincide, a decision about the most likely response can be made only upon examination of all the possibilities. Question marks indicate predictions that require measurement, because different pathways have opposite effects.

As noted in Chapter 3, it will not always be possible to identify the source of an impact. In such instances, however, correlations between variables (e.g., components) can still be examined. In model 14 (see Figure B.1), for example, N and P respond in the same direction to impacts entering the system through H_1, but respond in opposite directions to impacts entering through other modes. Therefore, a positive correlation between N and P as well as negative correlation between the variables and H_1 and H_2, identify the source of the impact as H_1. Table B.1 shows patterns of correlations among variables for each model and each input mode. There are four columns of such tables corresponding to the four possible sources of input heading the columns. The 1's along the diagonal identify variables that change in either direction; zero on the diagonal indicates variables that will not change regardless of the source of input.

No two tables are identical in the same horizontal row corresponding to a single model. If the model has been validated already, an examination of the correlation patterns identifies the source of input. Some tables are unique, identifying both source and model. Inputs entering through H_2, however, often have consequences that are insensitive to model differences. The methods used here are described in Levins (1975).

Uses of this approach for predicting impacts of chemical substances are reviewed in Chapter 3.

REFERENCE

Levins, R. (1975) Evolution in communities near equilibrium. Pages 16-50, Ecology and Evolution of Communities, edited by M.L. Cody and J.M. Diamond. Cambridge, Mass.: Harvard University Press.

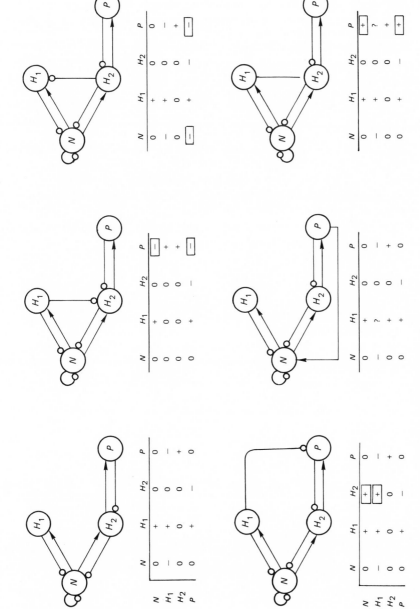

FIGURE B.1   Hypothetical models of ecosystem interactions.

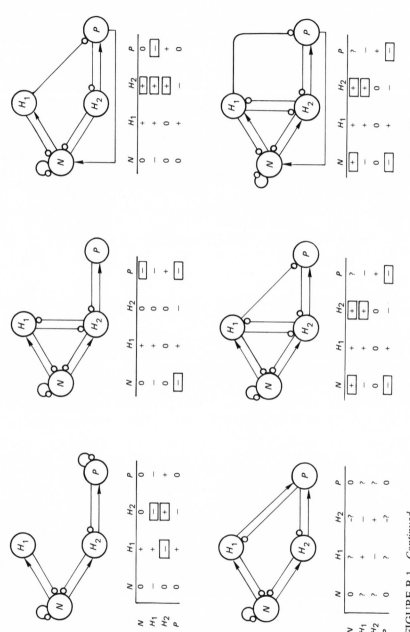

FIGURE B.1 *Continued.*

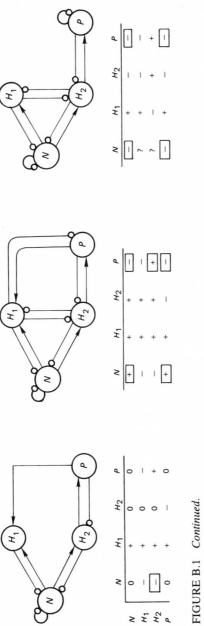

FIGURE B.1 *Continued.*

TABLE B.1 Correlation Patterns Among Ecosystem Variables

Model	N				H_1				H_2				P			
	N	H_1	H_2	P	N	H_1	H_2	P	N	H_1	H_2	P	N	H_1	H_2	P
1																
N	0				1				0				0			
H_1	0	1			−	−			0	0			0	1		
H_2	0	0	0		0	0	0		0	0	0		0	−	1	
P	0	0	0	0	+	−	0	1	0	0	0	1	0	0	0	0
2																
N	0				1				0				0			
H_1	0	1			−	−			0	0			0	1		
H_2	0	0	0		0	0	0		0	0	0		0	−	1	
P	0	−	0	1	+	−	0	1	0	0	0	1	0	+	−	1
3																
N	0				1				0				1			
H_1	0	1			−	−			0	0			−	1		
H_2	0	0	0		0	0	0		0	0	0		+	−	1	
P	0	0	0	0	+	−	0	1	0	0	0	1	0	0	0	0
4																
N	0				1				0				0			
H_1	0	1			−	−			0	0			0	1		
H_2	0	+	1		−	+	1		0	0	0		0	−	1	
P	0	0	0	0	+	−	−	1	0	0	0	1	0	0	0	0
5																
N	0				1				0				0			
H_1	0	1			−	−			0	1			0	1		
H_2	0	0	0		0	0	0		0	0	0		0	−	1	
P	0	0	0	0	+	−	0	1	0	+	0	1	0	0	0	0
6																
N	0				1				0				0			
H_1	0	1			−	−			0	0			0	+		
H_2	0	0	0		0	0	0		0	0	0		0	−	1	
P	0	+	0	1	?	?	0	1	0	0	0	1	0	0	0	0
7																
N	0				1				0				0			
H_1	0	1			−	−			0	1			0	1		

	H_2			0	0			1	+			1	+			1	0
	P				0				1				1				0
8	N	0	0	0	0	1	–	0	+	0	0	0	0	1	–	+	+
	H_1		1	0	–		1	0	–		0	0	0		1	–	–
	H_2			0	0			0	0			0	0			1	+
	P				1				1				1				1
9	N	0	0	0	0	1	–	–	+	0	0	0	0	0	0	0	0
	H_1		1	+	0		1	+	–		0	0	0		1	–	0
	H_2			1	0			1	–			1	+			1	0
	P				0				1				1				0
10	N	0	0	0	0	1	a	-a	?	1	a	-a	?	0	0	0	0
	H_1		1	–	0		1	–	?		1	–	b		1	–	0
	H_2			1	0			1	?			1	-b			1	0
	P				0				1				1				0
11.12	N	1	+	+	?	1	–	–	+	0	0	0	0	1	–	+	+
	H_1		1	+	?		1	+	–		0	0	0		1	–	–
	H_2			1	?			1	–			0	0			1	+
	P				1				1				1				1
13	N	0	0	0	0	1	–	0	+	1	–	0	–	0	0	0	0
	H_1		1	0	0		1	0	–		1	0	+		1	–	0
	H_2			0	0			0	0			0	0			1	0
	P				0				1				1				0
14	N	1	+	+	–	1	–	–	+	1	–	–	–	1	+	–	–
	H_1		1	+	–		1	+	–		1	+	+		1	–	–
	H_2			1	–			1	–			1	+			1	+
	P				1				1				1				1
15	N	1	–	+	+	1	a	-a	-a	1	a	-a	-a	1	–	+	+
	H_1		1	–	–		1	–	–		1	–	–		1	–	–
	H_2			1	+			1	+			1	+			1	+
	P				1				1				1				1